Biofuels

Wiley Series in Renewable Resources

Series Editor

Christian V. Stevens, Department of Organic Chemistry, Ghent University, Belgium

Titles in the Series

Forthcoming Titles

Biofuels

Edited by

WIM SOETAERT

Ghent University, Ghent, Belgium

ERICK J. VANDAMME

Ghent University, Ghent, Belgium

A John Wiley and Sons, Ltd., Publication

This edition first published 2009
© 2009 John Wiley & Sons, Ltd

Registered office
John Wiley & Sons, Ltd, The Atrium, Southern Gate, Chichester, West Sussex, PO19 8SQ, United Kingdom

For details of our global editorial offices, for customer services and for information about how to apply for permission to reuse the copyright material in this book please see our website at www.wiley.com.

The right of the author to be identified as the author of this work has been asserted in accordance with the Copyright, Designs and Patents Act 1988.

Reprinted April and October 2009

Wiley also publishes its books in a variety of electronic formats. Some content that appears in print may not be available in electronic books.

Designations used by companies to distinguish their products are often claimed as trademarks. All brand names and product names used in this book are trade names, service marks, trademarks or registered trademarks of their respective owners. The publisher is not associated with any product or vendor mentioned in this book. This publication is designed to provide accurate and authoritative information in regard to the subject matter covered. It is sold on the understanding that the publisher is not engaged in rendering professional services. If professional advice or other expert assistance is required, the services of a competent professional should be sought.

The publisher and the author make no representations or warranties with respect to the accuracy or completeness of the contents of this work and specifically disclaim all warranties, including without limitation any implied warranties of fitness for a particular purpose. This work is sold with the understanding that the publisher is not engaged in rendering professional services. The advice and strategies contained herein may not be suitable for every situation. In view of ongoing research, equipment modifications, changes in governmental regulations, and the constant flow of information relating to the use of experimental reagents, equipment, and devices, the reader is urged to review and evaluate the information provided in the package insert or instructions for each chemical, piece of equipment, reagent, or device for, among other things, any changes in the instructions or indication of usage and for added warnings and precautions. The fact that an organization or Website is referred to in this work as a citation and/or a potential source of further information does not mean that the author or the publisher endorses the information the organization or Website may provide or recommendations it may make. Further, readers should be aware that Internet Websites listed in this work may have changed or disappeared between when this work was written and when it is read. No warranty may be created or extended by any promotional statements for this work. Neither the publisher nor the author shall be liable for any damages arising herefrom.

Library of Congress Cataloging-in-Publication Data

Soetaert, Wim.
    Biofuels / Wim Soetaert, Erick J. Vandamme.
      p. cm. – (Wiley series in renewable resource)
    Includes bibliographical references and index.
    ISBN 978-0-470-02674-8 (cloth)
      1. Biomass energy–Technological innovations.   2. Biomass energy–Economic aspects.
    3. Renewable natural resources.   I. Vandamme, Erick J., 1943–   II. Title.
    TP339.S64   2008
    333.95′39–dc22   2008027967

A catalogue record for this book is available from the British Library.

ISBN 978-0-470-02674-8 (H/B)

Set in 10/12pt Times by Aptara Inc., New Delhi, India
Printed and bound in Great Britain by CPI Antony Rowe, Chippenham, Wiltshire

Contents

Series Preface

Renewable resources, their use and modification are involved in a multitude of important processes with a major influence on our everyday lives. Applications can be found in the energy sector, chemistry, pharmacy, the textile industry, paints and coatings, to name but a few.

The area interconnects several scientific disciplines (agriculture, biochemistry, chemistry, technology, environmental sciences, forestry, ...), which makes it very difficult to have an expert view on the complicated interaction. Therefore, the idea to create a series of scientific books, focussing on specific topics concerning renewable resources, has been very opportune and can help to clarify some of the underlying connections in this area.

In a very fast changing world, trends are not only characteristic for fashion and political standpoints, also science is not free from hypes and buzzwords. The use of renewable resources is again more important nowadays; however, it is not part of a hype or a fashion. As the lively discussions among scientists continue about how many years we will still be able to use fossil fuels, with opinions ranging from 50 years to 500 years, they do agree that the reserve is limited and that it is essential not only to search for new energy carriers but also for new material sources.

In this respect, renewable resources are a crucial area in the search for alternatives for fossil-based raw materials and energy. In the field of energy supply, biomass and renewable-based resources will be part of the solution alongside other alternatives such as solar energy, wind energy, hydraulic power, hydrogen technology and nuclear energy.

In the field of material sciences, the impact of renewable resources will probably be even bigger. Integral utilization of crops and the use of waste streams in certain industries will grow in importance, leading to a more sustainable way of producing materials.

Although our society was much more (almost exclusively) based on renewable resources centuries ago, this disappeared in the Western world in the nineteenth century. Now it is time to focus again on this field of research. However, it should not mean a *retour à la nature*, but it should be a multidisciplinary effort on a highly technological level to perform research towards new opportunities, to develop new crops and products from renewable resources. This will be essential to guarantee a level of comfort for a growing number of people living on our planet. It is 'the' challenge for the coming generations of scientists to develop more sustainable ways to create prosperity and to fight poverty and hunger in the world. A global approach is certainly favoured.

This challenge can only be dealt with if scientists are attracted to this area and are recognized for their efforts in this interdisciplinary field. It is therefore also essential that consumers recognize the fate of renewable resources in a number of products.

Furthermore, scientists do need to communicate and discuss the relevance of their work. The use and modification of renewable resources may not follow the path of the genetic engineering concept in view of consumer acceptance in Europe. Related to this aspect, the series will certainly help to increase the visibility of the importance of renewable resources.

Being convinced of the value of the renewables approach for the industrial world, as well as for developing countries, I was myself delighted to collaborate on this series of books focussing on different aspects of renewable resources. I hope that readers become aware of the complexity, the interaction and interconnections, and the challenges of this field and that they will help to communicate on the importance of renewable resources.

I certainly want to thank the people of Wiley from the Chichester office, especially David Hughes, Jenny Cossham and Lyn Roberts, in seeing the need for such a series of books on renewable resources, for initiating and supporting it and for helping to carry the project to the end.

Last, but not least I want to thank my family, especially my wife Hilde and children Paulien and Pieter-Jan for their patience and for giving me the time to work on the series when other activities seemed to be more inviting.

Christian V. Stevens
Faculty of Bioscience Engineering
Ghent University, Belgium
Series Editor '*Renewable Resources*'
June 2005

Preface

This volume on *Biofuels* fits within the series Renewable Resources. It covers the use and conversion technologies of biomass as a renewable resource to produce bio-energy in a sustainable way, mainly in the form of liquid and gaseous biofuels.

These biofuels are a convenient renewable energy carrier for specific purposes, with transportation as an important application sector. Renewable biomass is produced annually, based on photosynthesis, and is available in different forms, depending on climatic conditions and economic situations around the world. Chemical and thermochemical methods, as well as fermentation and biocatalysis technologies, are essential to efficiently convert biomass directly or indirectly into biofuels, with bio-ethanol, biodiesel and biogas as today's main practical players. In this context, green biotechnology, green chemistry and white biotechnology are to join forces to arrive at sustainable processes and fuels. The use of biofuels is quickly gaining momentum all over the world, and can be expected to have an ever-increasing impact on the energy and agricultural sector in particular. New and efficient 'bio-cracking' technologies for biomass are under development, while existing (thermo)chemical, fermentation and enzyme technologies are further optimized. These developments cover basic and applied research, pilot scale experimentation and demonstration plants for second generation biofuels.

All foregoing scientific and technological aspects are treated in this volume by renowned experts in their field. In addition, the economical and ecological aspects of biofuels development and application are receiving due attention: market developments are commented as well as the sustainability of biofuels production and use. Particularly, the links between the technical, economical and ecological aspects are clearly expressed in this volume and are actually covered here for the first time in a single comprehensive volume. The editors are indebted to the John Wiley & Sons staff (Jenny Cossham, Zoë Mills, Richard Davies) for their invaluable supportive help along the editorial process, and to the secretarial input of Dominique Delmeire (Ghent University) who kept us abreast of the 'labour' efforts of all the contributors. Without all of them, this volume would not have been born and grown into an active youngster, a real player in and on the biofuels-field.

Wim Soetaert
Erick J. Vandamme
Ghent, January 2008

Note on conversion factors

The following conversion factors can be used:

1 acre = 0.4047 hectare
1 US bushel of corn = 35.2 litres = 25.4 kg
1 US gallon = 3.78541 litre.

List of Contributors

Editors

Wim Soetaert   Laboratory of Industrial Microbiology and Biocatalysis, Faculty of Bioscience Engineering, Ghent University, Ghent, Belgium

Erick J. Vandamme   Laboratory of Industrial Microbiology and Biocatalysis, Faculty of Bioscience Engineering, Ghent University, Ghent, Belgium

Contributors

Matthew T. Carr   Policy Director, Industrial and Environmental Section, Biotechnology Industry Organization, Washington, USA.

Pieternel A.M. Claassen   Agrotechnology and Food Sciences group, Wageningen University and Research Center, Wageningen, The Netherlands.

Brent Erickson   Executive Vice President, Industrial & Environmental Section, Biotechnology Industry Organization, Washington, USA.

Hideki Fukuda   Division of Molecular Science, Graduate School of Science and Technology, Kobe University, Japan.

Paul Gallagher   Department of Economics, Iowa State University, Iowa, USA.

Heleen P. Goorissen   Laboratory of Microbiology, Wageningen University and Research Center, Wageningen, The Netherlands.

Adrianus Van Haandel, Federal University of Paraíba, Department of Civil Engineering, Campina Grande, Brazil.

James R. Hettenhaus   President and CEO, Chief Executive Assistance, Inc. Charlotte. NC, USA.

Barnim Jeschke   Co-founder and former Non-Executive Director, ELSBETT Technologies GmbH, Munich, Germany.

Servé W.M. Kengen   Laboratory of Microbiology, Wageningen University and Research Center, Wageningen, The Netherlands.

Martin Mittelbach   Department of Renewable Resources, Institute of Chemistry, Karl-Franzens-University, Graz, Austria.

Ed W.J. van Niel   Laboratory of Applied Microbiology, University of Lund, Sweden.

René van Ree   Wageningen University and Research Centre, Wageningen, The Netherlands.

Hosein Shapouri   USDA, OCE, OE, Washington, DC, USA.

Alfons J.M. Stams   Wageningen University and Research Centre, Wageningen, The Netherlands.

Christian V. Stevens   Faculty of Bioscience-engineering, Department of Organic Chemistry, Ghent University, Ghent, Belgium.

Marcel Verhaart   Laboratory of Microbiology, Wageningen University and Research Center, Wageningen, The Netherlands.

Roland Verhé   Faculty of Bioscience-engineering, Department of Organic Chemistry, Ghent University, Ghent, Belgium.

Willy Verstraete   Faculty of Bioscience-engineering, Laboratory of Microbial Ecology and Technology, Ghent University, Ghent, Belgium

Arnaldo Walter   Department of Energy and NIPE, State University of Campinas (Unicamp), Brazil.

Peter Weiland   Bundesforschungsanstalt für Landwirtschaft, Institut für Technologie und Biosystemtechnik, Braunschweig, Germany.

Robin Zwart   Energy Research Centre of the Netherlands Biomass, Coal and Environmental Research Petten, The Netherlands.

1

Biofuels in Perspective

W. Soetaert and Erick J. Vandamme

Laboratory of Industrial Microbiology and Biocatalysis, Faculty of Bioscience Engineering, Ghent University, Ghent, Belgium

1.1 Fossil versus Renewable Energy Resources

Serious geopolitical implications arise from the fact that our society is heavily dependent on only a few energy resources such as petroleum, mainly produced in politically unstable oil-producing countries and regions. Indeed, according to the World Energy Council, about 82 % of the world's energy needs are currently covered by fossil resources such as petroleum, natural gas and coal. Also ecological disadvantages have come into prominence as the use of fossil energy sources suffers a number of ill consequences for the environment, including the greenhouse gas emissions, air pollution, acid rain, etc. (Wuebbles and Jain, 2001; Soetaert and Vandamme, 2006).

Moreover, the supply of these fossil resources is inherently finite. It is generally agreed that we will be running out of petroleum within 50 years, natural gas within 65 years and coal in about 200 years at the present pace of consumption. With regard to the depletion of petroleum supplies, we are faced with the paradoxical situation that the world is using petroleum faster than ever before, and nevertheless the 'proven petroleum reserves' have more or less remained at the same level for 40 years, mainly as a result of new oil findings (Campbell, 1998). This fact is often used as an argument against the 'prophets of doom', as there is seemingly still plenty of petroleum around for the time being. However, those 'proven petroleum reserves' are increasingly found in places that are poorly accessible, inevitably resulting in an increase of extraction costs and hence, oil prices. Campbell and Laherrère (1998), well-known petroleum experts, have predicted that the world production

Biofuels Edited by Wim Soetaert and Erick J. Vandamme

of petroleum will soon reach its maximum production level (expected around 2010). From then on, the world production rate of petroleum will inevitably start decreasing.

As the demand for petroleum is soaring, particularly to satisfy economically skyrocketing countries such as China (by now already the second largest user of petroleum after the USA) and India, petroleum prices are expected to increase further sharply. The effect can already be seen today, with petroleum prices soaring to over 90 $/barrel at the time of writing (September 2007). Whereas petroleum will certainly not become exhausted from one day to another, it is clear that its price will tend to increase. This fundamental long-term upward trend may of course be temporarily broken by the effects of market disturbances, politically unstable situations or crises on a world scale.

Worldwide, questions arise concerning our future energy supply. There is a continual search for renewable energy sources that will in principle never run out, such as hydraulic energy, solar energy, wind energy, tidal energy, geothermal energy and also energy from renewable raw materials such as biomass. Wind energy is expected to contribute significantly in the short term (Anonymous, 1998). Giant windmill parks are already on stream and more are being planned and built on land and in the sea. In the long run, more input is expected from solar energy, for which there is still substantial technical progress to be made in the field of photovoltaic cell efficiency and production cost (Anonymous, 2004). Bio-energy, the renewable energy released from biomass, is expected to contribute significantly in the mid to long term. According to the International Energy Agency (IEA), bio-energy offers the possibility to meet 50 % of our world energy needs in the 21st century.

In contrast to fossil resources, agricultural raw materials such as wheat or corn have until recently been continuously declining in price because of the increasing agricultural yields, a tendency that is changing now, with competition for food use becoming an issue. New developments such as genetic engineering of crops and the production of bio-energy from agricultural waste can relieve these trends.

Agricultural crops such as corn, wheat and other cereals, sugar cane and beets, potatoes, tapioca, etc. can be processed in so-called biorefineries into relatively pure carbohydrate feedstocks, the primary raw material for most fermentation processes. These fermentation processes can convert those feedstocks into a wide variety of valuable products, including biofuels such as bio-ethanol.

Oilseeds such as soybeans, rapeseed (canola) and palm seeds (and also waste vegetal oils and animal fats), can be equally processed into oils that can be subsequently converted into biodiesel (Anonymous, 2000; Du et al., 2003). Agricultural co-products or waste such as straw, bran, corn cobs, corn stover, etc. are lignocellulosic materials that are now either poorly valorized or left to decay on the land. Agricultural crops or organic waste streams can also be efficiently converted into biogas and used for heat, power or electricity generation (Lissens et al., 2001). These raw materials attract increasing attention as an abundantly available and cheap renewable feedstock. Estimations from the US Department of Energy have shown that up to 500 million tonnes of such raw materials can be made available in the USA each year, at prices ranging between 20 and 50 $/ton (Clark 2004).

1.2   Economic Impact

For a growing number of technical applications, the economic picture favours renewable resources over fossil resources as a raw material (Okkerse and Van Bekkum, 1999). Whereas

Table 1.1 *Approximate average world market prices in 2007 of renewable and fossil feedstocks and intermediates*

Fossil		Renewable	
Petroleum	400 €/t	Corn	150 €/t
Coal	40 €/t	Straw	20 €/t
Ethylene	900 €/t	Sugar	250 €/t
Isopropanol	1000 €/t	Ethanol	500 €/t

this is already true for a considerable number of chemicals, increasingly produced from agricultural commodities instead of petroleum, this is also becoming a reality for the generation of energy. The prices given in Table 1.1 are the approximate average world market prices for 2007. Depending on local conditions such as distance to production site and local availability, these prices may vary rather widely from one place to another. Also, protectionism and local subsidies may seriously distort the price frame. As fossil and renewable resources are traded in vastly diverging measurement units and currencies, one needs to convert the barrels, bushels, dollars and euros into comparable units to turn some sense into it. All prices were converted into Euro per metric ton (dry weight) for a number of fossil or renewable raw material as well as important feedstock intermediates such as ethylene and sugar, for the sole purpose of a clear indicative cost comparison of fossil versus renewable resources.

From Table 1.1, one can easily deduce that on a dry weight basis, renewable agricultural resources cost about half as much as comparable fossil resources. Agricultural co-products such as straw are even a factor 10 cheaper than petroleum. At the present price of crude oil (> 90 $/barrel, corresponding to 400 €/t in September 2007), petroleum costs about three times the price of corn. It is also interesting to note that the cost of sugar, a highly refined very pure feedstock (> 99.5 % purity), is about the same as petroleum, a very crude and unrefined mixture of chemical substances. As the energy content of renewable resources is roughly half the value of comparable fossil raw materials, one can conclude that on an energy basis, fossil and renewable raw materials are about equal in price. Also volume wise, agricultural feedstocks and intermediates have production figures in the same order of magnitude as their fossil counterparts, as indicated in Table 1.2.

It is obvious that agricultural feedstocks are cheaper than their fossil counterparts today and are readily available in large quantities. What blocks their further use is not economics but the lack of appropriate conversion technology. Whereas the (petro)chemical technology base for converting fossil feedstocks into a bewildering variety of useful products is by now very efficient and mature, the technology for converting agricultural raw materials into chemicals, materials and energy is still in its infancy.

It is widely recognized that new technologies will need to be developed and optimized in order to harvest the benefits of the bio-based economy. Particularly industrial biotechnology is considered a very important technology in this respect, as it is excellently capable to use agricultural commodities as a feedstock (Demain, 2000, 2007; Dale 2003; Vandamme and Soetaert, 2004). The processing of agricultural feedstocks into useful products occurs in so-called biorefineries (Kamm and Kamm, 2004; Realff and Abbas, 2004). Whereas the gradual transition from a fossil-based society to a bio-based society will take time and

Table 1.2   *Estimated world production and prices for renewable feedstocks and petrochemical base products and intermediates*

	Estimated world production (million tons per year)	Indicative world market price (euro per ton)
Renewable feedstocks		
Cellulose	320	500
Sugar	140	250
Starch	55	250
Glucose	30	300
Petrochemical base products and intermediates		
Ethylene	85	900
Propylene	45	850
Benzene	23	800
Caprolactam	4	2000

effort, it is clear that renewable raw materials are going to win over fossil resources in the long run. This is particularly true in view of the perspective of increasingly rarer, difficult to extract and more expensive fossil resources.

1.3   Comparison of Bio-energy Sources

1.3.1   Direct Burning of Biomass

Traditional renewable biofuels, such as firewood, used to be our most important energy source and they still fulfill an important role in global energy supplies today. The use of these traditional renewable fuels covered in 2002 no less than 14.2 % of the global energy use, far more than the 6.9 % share of nuclear energy (IEA). In many developing countries, firewood is still the most important and locally available energy source, but equally so in industrialized countries. The importance is even increasing: in several European countries, new power stations using firewood, forestry residues or straw have recently been put into operation and there are plans to create energy plantations with fast growing trees or elephant grass (*Miscanthus* sp.). On the base of net energy generation per ha, such energy plantations are the most efficient process to convert solar energy through biomass into useful energy. An important factor in this respect is that such biofuels have (in Western Europe) high yields per ha (12 t/ha and more) and can be burnt directly, giving rise to an energy generation of around 200 GJ/ha/yr (Table 1.3).

1.3.2   Utilization Convenience of Biofuels

The energy content of an energy carrier is, however, only one aspect in the total comparison. For the value of an energy carrier is not only determined through its energy content and yield per hectare, but equally by its physical shape and convenience in use. This aspect of an energy source is particularly important for mobile applications, such as transportation. In Europe, the transport sector stands for 32 % of all energy consumption, making it a very

Table 1.3 *Energy yields of bio-energy crops in Flanders (Belgium)*

	Yield (t dry matter/ha/yr)	Biofuel-type	Biofuel yield t/ha/yr	Gross energy yield	
				GJ/t	GJ/ha/yr
Wheat (cereal)	6.8	bio-ethanol	2.29	26.8	61
Sugar beets (root)	14	bio-ethanol	4.84	26.8	130
Rapeseed (seed)	3.1	biodiesel	1.28	37	50
Willow/poplar (wood)	10.8	firewood	10.8	18	194

important energy user. There is consequently a strong case for the use of renewable fuels in the transport sector, particularly biofuels. Whereas in principle, we can drive a car on firewood, this approach is all but user friendly. In practice, liquid biofuels are much better suited for such an application. It is indeed no coincidence that nearly all cars and trucks are powered by liquid fuels such as gasoline and diesel. These liquid fuels are easily and reliably used in classic explosion engines and they are compact energy carriers, leading to a large action radius of the vehicle. They are easily stored, transported and transferred (it takes less than a minute to fill up your tank) and their use basically requires no storage technology at all (a simple plastic fuel tank is sufficient). Our current mobility concept is consequently mainly based on motor vehicles powered by liquid fuels that are supplied and distributed through tank stations.

The current strong interest in liquid motor fuels such as bio-ethanol and biodiesel based on renewable sources is based strongly on the fact that these biofuels show all the advantages of the classic (fossil-based) motor fuels. They are produced from agricultural raw materials and are compact, user-friendly motor fuels that can be mixed with normal petrol and diesel, with no engine adaptation required. The use of bio-ethanol or biodiesel therefore fits perfectly within the current concept of mobility. Current agricultural practices, such as the production of sugar cane or beets, rapeseed or cereals also remain fundamentally unaltered. The introduction of these energy carriers does not need any technology changes and the industrial processes for mass production of biofuels are also available.

Table 1.3 compares the energy yields of the different plant resources and technologies. For comparison, rapidly growing wood species such as willow or poplar as a classical renewable energy source, are also included.

It is clear that the gross energy yield per hectare is the highest for fast growing trees such as willow or poplar. However, a car does not run on firewood. Even if we restrict ourselves to the liquid fuels, there remain big differences between the different bio-energy options to be explained.

1.3.3 Energy Need for Biofuel Production

At first sight, based on gross energy yield per hectare, bio-ethanol from sugar beets would appear the big winner, combining a high yield per hectare and a high energy content of the produced bio-ethanol. Bio-ethanol out of wheat is lagging behind and biodiesel out of rapeseed comes last. Yet, biodiesel produced out of rapeseed is currently rapidly progressing in production volume, especially in Europe. The comparison is clearly more complex than would appear at first sight, with several other facts to be considered.

1.3.3.1   *Comparison of Biofuels to Fossil Fuels*

The energy input in the cultivation of the plants, transport as well as the production process itself needs to be taken into account. During the production of bio-ethanol, the distillation process is a big energy consumer. The amount of energy needed to produce the bio-ethanol is even close to the amount of energy obtained from the bio-ethanol itself. Shapouri et al. (2003) have carefully studied the energy balance of corn ethanol and have concluded that the energy output:input ratio is 1.34. When all energy inputs are taken into account, the net energy yield can even be negative in poorly efficient production processes. It would then appear that more energy is being used than is produced. Ironically, this energy input often comes out of fossil energy sources, except in Brazil, where renewable sugar cane bagasse contributes increasingly to the energy input. Obviously, this point is frequently used by opponents of bio-ethanol; they even consider it as an unproductive way to convert fossil energy in so called bio-energy, for the only sake of pleasing the agricultural sector.

Dale (2007) has nicely shown the inconsistency of the 'net energy' debate, by pointing at the reality that all energy sources are not equal. One unit of energy from petrol is e.g. much more useful than the same amount of energy in coal. Whereas net energy analysis is simple and has great intuitive appeal, it is also dangerously misleading. For making wise decisions about alternative fuels, we need to carefully choose our metrics of comparison. Dale suggests two complementary metrics as being far more sensible than net energy. First, alternative fuels (e.g. ethanol) can be rated on their ability to displace petroleum; and second, ethanol could be rated on the total greenhouse gases produced per km driven.

Sheehan et al. (2004) have determined the Fossil Energy Replacement Ratio (FER), the energy delivered to the customer over the fossil energy used. This parameter is important in relation to the emission of carbon dioxide, the most important greenhouse gas. A high FER means that less greenhouse gases are produced (from the fossil fuel input) per unit of energy delivered to the customer. They have found a FER of 1.4 for bio-ethanol based on corn, and a FER as high as 5.3 for bio-ethanol based on lignocellulosic raw materials such as straw or corn stover. For comparison, the FER for gasoline is 0.8 and for electricity it is as low as 0.4.

In order to properly evaluate this development, one must also consider that bio-ethanol is a high-quality and energy-dense liquid fuel, perfectly usable for road transport. For its production, one needs mainly energy in the form of heat (for distillation), a fairly cheap, low-quality and non-portable energy source. The conversion of one energy form into another, especially if it becomes portable, is indeed a productive process. In the case of biofuel production, one converts cheap low-quality heat and biomass into high-quality portable liquid motor fuel, a relatively expensive but very convenient source of energy, particularly for transportation use. In the same way, cars do not run on petroleum either, but on the fuel that is being distilled out of it. The distillation, extraction and long-distance transport of petroleum also require a large energy input. The matter of the fossil energy-input into bio-ethanol production becomes a non-issue altogether, when biomass is used as the source of heat, as is commonly practised in Brazil where the sugar cane residue bagasse is burnt to generate the heat required for distillation. Similar production schedules may soon become a reality in the USA or Europe when e.g. ethanol is produced from wheat, with the wheat straw being burned to generate the heat for distillation.

Biodiesel has a lower energy input required for its production. However, the low yield of the crops from which it is produced per hectare dampens the perspectives for biodiesel. Concerning the difference in gross energy yield per hectare between sugar beet and wheat, it also has to be borne in mind that European farmers have traditionally obtained very high prices for their sugar beets. This high price was being maintained by the sugar market regulation (quota regulation) in Europe, which is now under reform. Even if the yield per hectare is higher for sugar beets, with the current price structure it is today more economical to produce ethanol out of wheat or other cereals, unless the ethanol production can be coupled to the sugar production, a production scheme that offers technical advantages.

1.4   Conclusion

The use of bio-ethanol and biodiesel derived from agricultural crops is a technically viable alternative for fossil-based gasoline or diesel. Moreover, their use fits perfectly in the present concept and technology of our mobility. Liquid energy carriers are an (energetically) expensive but very useful energy carrier for mobile applications such as transportation. It is clear that energy sources for mobile applications should not only be compared on the basis of simple energy balances or costs, but also on the base of their practical usefulness, quality, environmental characteristics and convenience in use of the obtained energy carrier. It is interesting to note that Henry Ford, when designing his famous model T car, presumed that ethanol would become the car fuel of the future. Although initially petrochemistry got the upper hand, it now seems as though Henry Ford was way ahead of his time and proven right in the long run. Even as the discussion about the sense or nonsense of biofuels is ongoing, the transition process from a fossil-based to a bio-based society is clearly moving forward, with impressive growth in the USA, Brazil, China and Europe finally catching on. There is little doubt that in the medium term, we will all fill up our car with a considerable percentage of biofuels, probably unaware of it and without noticing any difference.

The large-scale introduction of biofuels can reconcile the interests of environment, mobility and agriculture and can be seen as an important step with high symbolic value towards the sustainable society of the future.

References

Anonymous (1998) Renewable energy target for Europe: 20 % by 2020. Report from the European Renewable Energy Council (EREC).

Anonymous (2000) Biodiesel: The Clean, Green Fuel for Diesel Engines (Fact sheet). NREL Report No. FS-580-28315; DOE/GO-102000-1048.

Anonymous (2004) Renewable Energy scenario to 2040: Half of the Global Energy Supply from renewables in 2040. Report from the European Renewable Energy Council (EREC).

Campbell, C.J. (1998) The future of oil. *Energy Explor. Exploit.* **16**: 125–52.

Campbell, C.J. and Laherrère, J.H. (1998) The end of cheap oil. *Sci. Am.* **278**: 78–83.

Clark, W. (2004) The case for biorefining. National Renewable Energy Laboratory. www.sae.org/
events/sfl/pres-clark.pdf

Dale, B.E. (2007) Thinking clearly about biofuels: ending the irrelevant net energy debate and developing better performance metrics for alternative fuels. *Biofpr, Biofuels, Bioproducts, Biorefining* **1**: 14–17.

Demain, A.L. (2000) Small bugs, big business: the economic power of the microbe. *Biotechnol. Adv.* **18**: 499–514.

Demain, A.L. (2007) The business of industrial biotechnology. *Industrial Biotechnology* **3**(3): 269–83.

Dale, B.E. (2003) 'Greening' the chemical industry: research and development priorities for biobased industrial products. *J. Chem. Technol. Biotechnol.* **78**: 1093–1103.

Du, W., Xu, Y. and Liu, D. (2003) Lipase-catalysed transesterification of soy bean oil for biodiesel production during continuous batch operation. *Biotechnol. Appl. Biochem.* **38**: 103–6.

IEA Bioenergy. www.ieabioenergy.com

Kamm, B. and Kamm, M. (2004) Principles of biorefineries. *Appl. Microbiol. Biotechnol.* **64**: 137–45.

Lissens, G., Vandevivere, P., De Baere, L., Biey, E.M., Verstraete, W. (2001). Solid waste digestors: process performance and practice for municipal solid waste digestion. *Water Science and Technology* **44**: 91–102.

Okkerse, C. and Van Bekkum, H. (1999) From fossil to green. *Green chem.* **1**(2): 107–14.

Realff, M.A. and Abbas, C. (2004) Industrial symbiosis : refining the biorefinery. *J. Industrial Ecology* **7**: 5–9.

Shapouri, H., Duffield, J.A. and Wang, M. (2003) The energy balance of corn ethanol: an update. USDA report no. 814.

Sheehan, J., Aden, A., Paustian, K., Killian, K., Brenner, J. et al. (2004) Energy and environmental aspects of using corn stover for fuel ethanol. *J. Industrial Ecology* **7**: 117–46. NREL Report No. JA-510-36462.

Soetaert, W., and Vandamme, E.J. (2006) The impact of industrial biotechnology. *Biotechnology. J.* **1**(7–8): 756–69.

Vandamme, E.J. and Soetaert, W. (2004) Industrial biotechnology and sustainable chemistry. BACAS-report, Royal Belgian Academy, 34p. http://www.europabio.org/documents/150104/bacas_report_en.pdf

Wuebbles, D.J. and Jain, A.K. (2001) Concerns about climate change and the role of fossil fuel use. *Fuel Process. Technol.* **71**: 99–119.

2

Sustainable Production of Cellulosic Feedstock for Biorefineries in the USA

Matthew T. Carr

Policy Director, Industrial and Environmental Section, Biotechnology Industry Organization, Washington, USA

James R. Hettenhaus

President and CEO, Chief Executive Assistance, Inc., Charlotte, NC, USA

2.1 Introduction

Demand for alternative feedstocks for fuels, chemicals and a range of commercial products has grown dramatically in the early years of the 21st century, driven by the high price of crude oil, government policy to promote alternatives and reduce dependence on foreign petroleum, and efforts to reduce net emissions of carbon dioxide and other greenhouse gases. This is particularly true for renewable feedstocks from agricultural sources.

For example, in the United States, ethanol production, primarily from corn grain, has more than tripled since 2000. Annual US production of ethanol is expected to exceed 7 billion gallons in 2008, displacing nearly 5 % of the projected 145 billion gallons of US gasoline demand.[1] Sales of biobased plastics are also expanding.

The growing availability of economically competitive biobased alternatives to petroleum can be attributed in large part to advances in the production and processing of corn grain for

Biofuels Edited by Wim Soetaert and Erick J. Vandamme
© 2009 John Wiley & Sons, Ltd

industrial uses. Steady increases in corn yields made possible by agricultural biotechnology continue to expand the supply of available feedstock, while rapid advances in the relatively new field of industrial biotechnology – including development of genetically enhanced microorganisms (GEMs) and specialized industrial enzymes – have greatly enhanced the efficiency of ethanol production.

Industrial biotechnology has also yielded a range of new biobased polymers, plastics and textiles. The US Department of Energy (DOE) has identified 12 building block chemicals that can be produced from biomass and converted to an array of high-value products.[2]

The National Corn Growers Association projects that with continued advances in biotechnology that boost corn yield, as much as 5.95 billion bushels of US grain could be available for ethanol and biobased products by 2015 – while continuing to satisfy food, animal feed and export demands. That amount of corn could produce nearly 18 billion gallons of ethanol, enough to meet over 10 % of projected US gasoline demand.[3]

But if ethanol is to expand into becoming a more widely available alternative to gasoline, new feedstock sources will be required in addition to high-efficiency production from grain. A robust sustainable supply chain for cellulosic biomass – biological material composed primarily of cellulose, such as agricultural and forestry residues, grasses, even municipal solid waste – is needed.

A recent comprehensive analysis by DOE and the US Department of Agriculture (USDA)[4] found that 'in the context of the time required to scale up to a large-scale biorefinery industry, an annual biomass supply of more than 1.3 billion dry tons can be accomplished'. Nearly one billion dry tons of this could be produced by American farmers, enough to meet the DOE goal of 60 billion gallons of ethanol production and 30 % displacement of petroleum by 2030.[5]

Recent advances in enzymes for the conversion of cellulosic biomass to sugars have brought ethanol from cellulose to the brink of commercial reality. A number of potential producers have announced plans to begin construction of cellulose-processing biorefineries in 2008.

One challenge for the emerging cellulosic biomass industry will be how to produce, harvest, store and deliver large quantities of feedstock to biorefineries in an economically and environmentally sustainable way. Farmers need up-to-date information on the effects of biomass removal to establish a better basis for sustainable collection, since commercial development of biorefineries may occur more quickly than previously believed. An evolution in crop-tilling practices toward no-till cropping will likely be needed in order to maintain soil quality while supplying adequate feedstock to these biorefineries. (No-till cropping is increasingly practiced but not yet widely utilized in regions of the country with the greatest potential to supply biomass.)

Additional infrastructure in collection, storage and transportation of biomass are also needed, including equipment for one-pass harvesting and investments in alternatives to trucking, such as short line rail. Further complicating matters is the absence of a clear protocol for pre-processing of cellulosic materials.

But sustainable production, harvest and processing of cellulosic biomass is achievable. Much of the future supply demand can be met by harvesting and utilizing residues from existing crops of corn, wheat, rice and other small grains. Production, collection and processing of these residues will deliver substantial economic and environmental benefits, including significant job creation in rural communities and mitigation of US emissions of greenhouse gases.

2.2 Availability of Cellulosic Feedstocks

A large, reliable, economic and sustainable feedstock supply is required for a biorefinery. Current yields for ethanol from agricultural residues (corn stover, straw from wheat, rice and other cereals, and sugarcane bagasse) are about 65 gallons per dry ton.[6] Thus, a moderately sized 65 million-gallon-per-year cellulosic biorefinery would need 1 million dry tons per year of feedstock. This could require 500,000 acres or more of cropland – a supply radius of at least 15 miles. The supply radius varies from 15 to 30 or more miles, depending on crop rotation, tillage practices, soil characteristics, topography, weather and farmer participation.

Research at a variety of sites indicates that economic delivery of crop residues is achievable at this radius and beyond – up to 50 miles from the biorefinery site when short line rail transport is available.[7] So, cellulosic biorefineries of well over 100 million gallon capacity are possible.

To sustain a commercial-scale biorefinery, cropland surrounding the site should meet the following criteria:

- large area: minimum of 500,000 acres of available cropland;
- sustainable: cropping practice maintains or enhances long-term health of the soil;
- reliable: consistent crop supply history with dry harvest weather;
- economic: high-yielding cropland;
- favorable transport: easy access from field to storage and processing facilities.

The recent USDA/DOE study on the technical feasibility of a billion-ton annual supply of biomass for bioenergy and biobased products estimated the potential amount of biomass available on an annual basis from agricultural sources in the United States at nearly 1 billion dry tons. Crop residues are the largest anticipated source. Assuming continued strong increases in corn yields from agricultural biotechnology and conversion of present cropping methods to no-till harvest (which allows for greater residue collection), the report estimates that 428 million dry tons of crop residues could be available on an annual basis by 2030. Most of the remainder, 377 million dry tons, is expected to come from new perennial energy crops.[4] The report anticipates the addition of 60 million acres of perennial energy crops as a market develops for cellulosic biomass. The development of high-yielding dedicated energy crops will be a critical element in achieving the DOE goal of 30 % petroleum displacement.

Of greater interest in the near term (3–5 years) is the current sustainable availability of biomass from agricultural lands. Corn stover – the leaves and stalks of the corn plant that remain after grain harvest (see Figure 2.1) – is the dominant near-term source of agricultural cellulosic biomass (see Figure 2.2), with substantial contributions from wheat straw, other small grain straw, soybeans and corn fiber. These figures assume a delivered price at the biorefinery of $30 per dry ton.

Figure 2.1 *Stover consists of the stalks, cobs and leaves that are usually left on the ground following corn harvest. Equipment for collection of corn stover must be developed, since few commercial uses for stover currently exist.*

Source: USDA Agricultural Research Service.

USDA/DOE Estimated Current Sustainable Availability of Cellulosic Biomass from Agricultural Lands

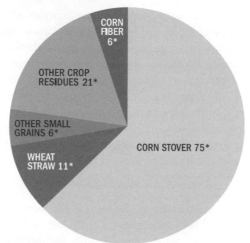

CORN FIBER 6*

OTHER CROP RESIDUES 21*

OTHER SMALL GRAINS 6*

WHEAT STRAW 11*

CORN STOVER 75*

* Figures above represent millions of dry tons per year.

Figure 2.2 *Cellulosic biomass currently available for sustainable collection in the US, according to a USDA-DOE analysis. Recent farmer colloquies suggest the actual available amount could be more than double this estimate.*

Source: Perlack, Wright et al., 2005.

In colloquies with farmers, potential processors and other stakeholders conducted by DOE during 2001 and again in 2003, there was general agreement that corn stover and cereal straw are the most likely near-term feedstocks for commercial-scale production of ethanol from cellulosic biomass.[8] However, farmers participating in the six Feedstock Roadmap Colloquies stated that the minimum price farmers would accept to collect biomass would be $50 per dry ton, or a return of at least $20 per acre net margin.[9]

At $50 per dry ton, the amount of economically recoverable, sustainably available biomass is more than double the amount estimated in the USDA/DOE report. Over 200 million dry tons of corn stover alone could be collected, enough to triple current ethanol production.

Future availability of feedstocks will depend on several variables, including crop acreage planted to meet competing demands; continued improvements derived from agricultural biotechnology; cropping practices and soil-quality maintenance considerations; and state and federal farm and energy policies. Coordination of farm and energy policies at both state and federal levels can serve to incentivize production, harvest and delivery of a variety of feedstocks to biorefineries.

As crop markets change – due to changing demand for food, animal feed, exports, and fuel and consumer products – farmers can be expected to adjust crop planting strategies to maximize their returns. As biorefinery construction creates markets for crop residues, farmers will have to adopt practices that lead to economic and sustainable removal. New models for maintaining soil quality also will be needed. Models based on soil organic material are currently in development.

2.3   Feedstock Options

Corn stover and cereal straw make up more than 80 % of currently available residues under both the USDA/DOE analysis and the $50 per dry ton scenario. Corn is the largest grain crop in the United States. Currently, 50 % of the corn biomass, about 250 million dry tons, is left in the field after harvest. Most of the available cereal straw biomass is from wheat (see Figure 2.3). Rice is also an important source, particularly in Texas and California. Sorghum, barley and oats have smaller potential.

There are significant regional differences in crop characteristics to consider, as well as differences in harvesting mechanics for stover and straw. More corn stover is available than straw, but straw is more readily removed (although in some areas it must be left in the field to retain moisture in the soil). Straw collection infrastructure is generally well developed, while corn stover collection is not. When cereal grain is ready to harvest, straw usually contains 20 % moisture or less, suitable for baling. In contrast, stover contains 50 % moisture and must remain in the field to dry and be collected later, depending on the weather. A wet harvest season can prevent its collection entirely.

Corn stover yields are 3–5 times greater – or more – on a per acre basis than straw from cereal crops. Unless cereal crops are irrigated, there is little straw left to collect. For example, the average dry land wheat straw yield is between 40 and 45 bushels per acre compared to 140–200 bushels per acre or more for corn stover. The equivalent of 20 bushels of straw must be left on the surface to comply with erosion guidelines with no-till. The

Figure 2.3 *Baling and collection technology for wheat straw has already been developed for a variety of commercial uses, such as animal bedding, landscape mulch, erosion control, and as a building material.*
Source: USDA Agricultural Research Service.

excess is less than 1 ton of straw per acre. In contrast, leaving 40 bushels of stover with no-till is often sufficient and the excess is 4 dry tons or more of stover per acre.

Soybean stubble is the surface material left after harvesting of the soy beans. Soybean stubble provides roughly the same feedstock quantity per acre as straw from dryland cereal grains. Little has published about its removal. More than 60 % of current soybean acres are no-till, and stubble availability could be considerably larger than straw, especially when a cover crop is included in the rotation to maintain soil quality. Alternatively, stubble availability could be negligible if high corn stover yields drive farmers towards adoption of continuous corn. The availability of soybean stubble will depend on the extent to which soybeans are used in rotation with corn, and the extent to which stubble is available under future tilling practice.

Bagasse presently offers limited opportunities as a feedstock in the United States. Bagasse is the remainder of the sugar cane plant after the sucrose is extracted at the sugar mill. Bagasse is currently burned, often inefficiently, to meet the energy needs of the sugar mill. Efficiency improvements in the burning process could reduce the amount of bagasse needed to power the processing plant by about one third, making excess bagasse available for fuel and chemical production. Production of fuels and chemicals from bagasse would also likely prove more profitable than simply burning it, so an even greater quantity may become available.

However, currently just 6 million dry tons of bagasse is produced in the United States. Even if burned efficiently, only enough for several fuel or chemical plants would be available. Much more cane could be grown if a market for the sugar existed or if the economics for conversion to fermentation sugars were demonstrated.

More bang from Bagasse

A high-fiber cultivar is under development in California. Switching to a high-fiber cane that is not suitable for sugar extraction but better for biomass conversion may open up considerable opportunity for growers. The high-fiber cane triples the cellulosic biomass available, to 110 tons per acre. Since the higher fiber content decreases the sucrose yield, it only becomes attractive when the bagasse can be processed to higher-value products. Bagasse has a composition close to corn stover. It is thought to have similar pretreatment and hydrolysis processing characteristics.

Corn fiber is being processed on a pilot basis now by several companies, including Aventine Renewable Energy, Inc. (formerly Williams BioEnergy), Broin, Abengoa and ADM. Their efforts are partially funded by DOE.

Corn fiber is a component of DDGs, the co-product of corn dry mill ethanol operations. It is a significant source of cellulose (see Table 2.1) Because corn fiber is already collected and delivered to ethanol facilities today, it represents a unique opportunity for cellulosic biorefining, since no additional collection or transportation infrastructure is needed. It could also provide an opportunity for farmer co-ops and other participants in grain ethanol production to participate in ethanol production from cellulose.

The biotechnology for corn fiber processing could eventually be applied to corn stover as well, though significant differences exist in the composition, consistency and price of the material. Corn fiber contains a small amount of lignin and a large amount of bound starch, while stover contains a much larger lignin fraction.

Table 2.1   *Corn fiber and stover composition, dry basis*

	Corn fiber	Stover
Cellulose	12 to 18%	32 to 38%
Hemicellulose	40 to 53%	28 to 32%
Lignin (Phenolic)	0.1 to 1%	15 to 17%
Starch	11 to 22%	None

Source: J. Hettenhaus.

Process waste from other sources, such as cotton gin trash and paper mill sludge, constitutes an additional potential source of cellulosic residues, especially for niche situations. However, volumes are small and, as with corn fiber, there is no consensus on whether these materials could provide an adequate supply of biomass to warrant biorefinery construction.

Dedicated energy crops include herbaceous perennials such as switchgrass (see Figure 2.4), other native prairie grasses and non-native grasses such as Miscanthus (see Figure 2.5), and short-rotation woody crops such as hybrid poplar and willow. There are currently no dedicated energy crops in commercial production, but the high biomass yield of such crops holds tremendous promise. Annual yields in excess of 8 dry tons per acre have already been achieved for both herbaceous and woody crops across a wide variety of conditions, with double this yield in some locations.[4]

The DOE and USDA anticipate that as many as 60 million acres of cropland, cropland pasture, and conservation acreage will be converted to perennial crop production once the technology for converting cellulosic biomass to ethanol is demonstrated at a commercial scale.

2.4 Sustainable Removal

Cellulosic biomass has the potential to revolutionize traditionally fossil-based industries, radically improving their environmental profile while revitalizing rural economies and enabling energy independence. This vision is only achievable if feedstocks are sustainably produced, harvested and processed.

Farm income expansion is possible only if crops can be grown and harvested without large amounts of fertilizer and other costly inputs. Soil quality enhancement, runoff reduction, greenhouse gas amelioration and other environmental benefits can be achieved with careful attention to production practices. Energy security gains depend on efficient collection, transport and processing of feedstocks. Each of these considerations will vary from region to region, even from farm to farm. Sustainable production practices must be tailored to each operation.

The availability of excess stover and straw for harvesting after erosion requirements are met is dependent on cropping practice and relative economic and environmental benefits. Tillage practice greatly affects availability. No-till practice allows most of the residue to be removed, especially when cover crops are employed.[10] In contrast, conventional tillage

Figure 2.4 *Switchgrass is a native perennial once found throughout the US Midwest and Great Plains. Research is underway to develop switchgrass as a dedicated energy crop.*

Source: USDA Agricultural Research Service.

leaves less than 30 % of the surface covered, and there is no excess residue available to remove. Since less than 20 % of farmers no-till and more than 60 % conventional till, a major shift in practice is needed for sustainable removal.

Sustainable delivery of cellulosic biomass feedstocks requires production and collection practices that do not substantially deplete the soil, such that large quantities of biomass may be harvested over sustained periods without sacrificing future yields.

Crop residues serve to both secure soil from erosion and restore nutrients to the soil through decomposition. With biomass removal, there is the potential for degradation of soil quality and increased erosion. From the perspective of soil and environmental quality, determining the amount of excess crop residue available for removal is a complex issue that will vary for different soils and management systems.

Figure 2.5 *Researchers at the University of Illinois at Urbana-Champaign are studying production and harvesting of the perennial grass Miscanthus as a dedicated energy crop.*
Source: John Caveny, photo courtesy of the University of Illinois.

An environmental and economic 'optimum' removal balances sufficient retention of residues to avoid erosion losses and maintain soil quality while using excess residue as biomass feedstock. The impact of varying levels of stover and straw removal will depend on local conditions and practices. Farmer involvement in the development of residue collection plans will be critical.

2.5   Erosion Control

Past studies of removal effects are helpful, especially for erosion control, but are often incomplete when addressing field removal of crop residues.[11] This is due partly to the wide variation in local conditions and system complexity – it is not an easy task – and partly to skepticism of the need for these studies. Several early attempts at removing biomass for industrial uses failed, and many potential participants remain concerned about soil tilth.

Excess availability of crop residue is dependent on the amount that must remain as soil cover to limit wind and water erosion. Erosion is a function of climate, soil properties, topography and cropping and support practices like contour planting and minimum tillage.

Figure 2.6 *Wind erosion of soil must be controlled for sustainable collection of biomass, particularly in the western United States.*
Source: USDA Agricultural Research Service.

Water erosion is of greatest concern in the eastern Corn Belt. Wind erosion becomes serious further west (see Figure 2.6)

2.6   Tilling Practice

A transition to conservation tillage practices, in which crops are grown with minimal cultivation of the soil, has been a key element of efforts to encourage more sustainable production of annual crops such as corn and wheat.

Under conventional tillage practices, where soils are intensively tilled to control weeds, deliver soil amendments and aid irrigation, less than 30 % of the soil is left undisturbed. All residues must be left on the field to prevent soil erosion, leaving no material available for collection.

With conservation tillage, 30 % or more of the soil is left covered. Some residue removal may be possible without threatening erosion control. No-till cropping, in which 100 % of the soil is left covered, allows for significant harvest of crop residues (see Figures 2.7 and 2.8). Approximately twice as much residue can be collected under no-till than under partial-till conservation practices.

The impact of tillage practice on feedstock availability for three different plant siting studies is shown in Table 2.2. Feedstock availability under current tilling practice and anticipated feedstock availability under no-till cropping are determined for each site using USDA Natural Resources Conservation Service (NRCS) erosion models.

Figure 2.7 *Conventional continuous-till cropping exposes soil to wind and water erosion. No-till cropping leaves soil covered, allowing for removal of substantial amounts of cellulosic biomass without damaging soil.*

Source: USDA Agricultural Research Service.

Figure 2.8 *Roots and stubble are left undisturbed with no-till cropping, securing soil.*

Source: USDA Agricultural Research Service.

Table 2.2   Comparison of feedstock production and cellulosic biomass available for collection within a 50-mile radius of three test sites under current tilling practice vs. no-till (millions of dry tons)

| Site study | Feedstock produced | Available for sustainable collection | |
		Under current tilling practice	With no-till
1. Wheat and sorghum, dry land	5.4	0	2.1
2. Corn Belt, dry land	5.4	1.8	3.6
3. Corn Belt, 50% irrigated	5.4	0.6	3.6

Source: J. Hettenhaus.

All three sites produce the same amount of crop residue according to USDA crop reports. With no-till, all sites could comfortably supply a 1 million-dry-ton biorefinery while complying with erosion guidelines. At the dry land wheat and sorghum site, which featured highly erodible soil, 40 % of the total residue, 2.1 million dry tons, was available for harvest under no-till cropping. Under current practice for this site, which is nearly all conventional till, no crop residue can be removed.

More stable soils provided 3.6 million dry tons of harvestable residues with no-till at both Corn Belt sites. Current practice reduced the available biomass by 50 % at the dry land site and by 83 % at the irrigated site. Corn-bean rotation at the dry land site allowed for greater collection under current practice than the irrigated site, which had more continuous corn with conventional tillage on irrigated acres.

Thus, under a range of conditions, no-till cropping allows for substantially greater residue collection than current practice, enabling biorefinery siting in areas where suitable supplies are currently unavailable.

Soil model limitations

It should be noted that soil erosion models have their limitations. They only indicate if soil is moved, not whether it is removed from a field. The models also do not provide a measure of soil quality. When residue is removed, reduced inputs from the residue to the soil can result in a negative flux from the soil and a loss of soil organic matter and other nutrients, leading to a breakdown of soil structure. Other models are under development to better measure soil quality, but are not expected to replace actual field measurements for some time. Managing for soil carbon quality helps ensure sustainable removal. The Soil Quality Index is recommended: http://csltest.ait.iastate.edu/SoilQualityWebsite/home.htm

2.7 Transitioning to No-till

Currently, more than half of land planted with corn, wheat and other cereals is under conventional tillage, and thus unavailable for residue collection. No-till cropping is practiced on less than 20 % of current acreage. To realize the full potential of cellulosic agricultural biomass, a significant evolution in cropping practices will be required.

Figure 2.9 shows adoption rates of no-till cropping for wheat, rice and corn from the most recent analysis by the Conservation Technology Information Center (CTIC) (http://www.conservationinformation.org/).[12] No-till cropping comprises less than 20 % of acreage in most counties throughout the country for each of these crops. But large regions with higher adoption rates exist, especially for spring wheat and corn.

The balance between conservation tillage and conventional tillage has remained relatively unchanged over the past decade, with roughly two-thirds of wheat acreage and 60 % of corn acreage under conventional tillage. Conventional tillage has been used on over 80 % of rice acreage since data collection began in 2000.

Figure 2.10 show the crop tilling history for wheat, rice and corn based on CTIC surveys. No-till remains a niche practice for rice, but there is a clear gradual evolution towards greater adoption of no-till for wheat and corn. No-till cropping has proven viable under a range of conditions for both crops. The success of no-till early adopters has prompted neighboring farmers to move to no-till, helping to form the localized regions of enhanced adoption seen in Figure 2.9.

No-till cropping also tends to reduce fuel and fertilizer use, substantially reducing operating cost. NRCS estimates that no-till cropping saves farmers an average of 3.5 gallons per acre in diesel fuel – an annual savings to farmers of about $500 million.[13] Recent price increases for fuel and fertilizer are expected to drive an even greater transition to no-till.

The local climate is a significant factor in considering crop residue removal, and the viability of no-till cropping will depend significantly on local conditions. In more arid regions surface cover is required for moisture retention in the soil. Thus, even with no-till cropping, the amount of residue available for collection in arid regions may be limited. But in wet regions, especially in the northern parts of the Corn Belt, collection of excess stover is desirable, since cooler soils under residues can delay or hamper crop germination and reduce yield. For example, farmers in the eastern corn belt have encountered problems with cool, moist soil conditions fostered by no-till's heavy residue cover.

Ultimately, demand for residues will likely prove a strong additional driver for the transition to no-till cropping. Once a market for agricultural residues develops, individual farmers or groups of farmers may elect to adopt no-till cropping to attract biorefineries to the area. Or, as Iogen has done with farmers in Idaho,[14] potential biorefinery project developers may seek out productive farmland and sign supply contracts that could require farmers to adopt no-till practices.

For dedicated energy crops such as switchgrass, tilling is not required on an annual basis, so soil quality maintenance is less of a concern. Most dedicated energy crops are perennial, requiring minimal tillage. Water and wildlife management are likely to be the primary environmental issues. These issues are addressed in considerable detail in a recent report from the Worldwatch Institute.[15]

2.8   Realizing Removal

In addition to production challenges, additional infrastructure in collection, storage and transportation is needed to supply a biorefinery. Farmers have accumulated considerable information on the impact of removing straw and corn stover on their farms and delivering it

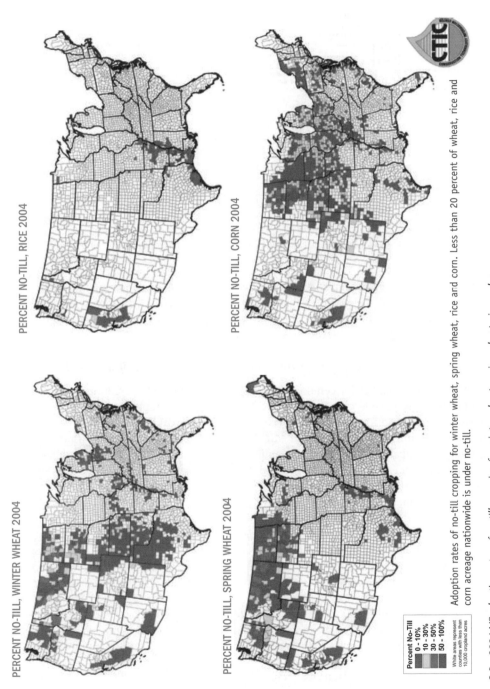

PERCENT NO-TILL, RICE 2004

PERCENT NO-TILL, CORN 2004

PERCENT NO-TILL, WINTER WHEAT 2004

PERCENT NO-TILL, SPRING WHEAT 2004

Percent No-Till
0 - 10%
10 - 30%
30 - 50%
50 - 100%

White areas represent
counties with less than
10,000 cropland acres

Adoption rates of no-till cropping for winter wheat, spring wheat, rice and corn. Less than 20 percent of wheat, rice and corn acreage nationwide is under no-till.

Figure 2.9 *2004 US adoption rates of no-till cropping for winter wheat, spring wheat, rice and corn.*

Source: Biotechnology Industry Organization; Conservation Technology Information Center.

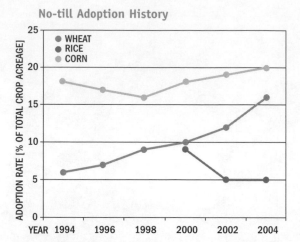

Figure 2.10   *Historic US adoption rates of no-till cropping for wheat, rice and corn.*
Source: Biotechnology Industry Organization; Conservation Technology Information Center.

to a processor. But for the most part this knowledge remains with the farmers, as no outside agencies were involved. Collection of feedstock on the proposed scale for biorefineries – as much as 30 times larger than those studied – will require a large, capable organization with considerable logistical expertise.

> **Sustainable collection case study: Imperial Young Farmers and Ranchers Project**
>
> With no current market for cellulosic biomass, identifying and overcoming potential obstacles to sustainable collection and delivery is a considerable challenge. But the Young Farmers and Ranchers of Imperial, Neb., have embarked on a study to do just that. With $3 million in funding from USDA and other sources, the Young Farmers are actively experimenting with innovative collection, pre-processing, storage and transport technologies for corn stover to identify logistical challenges and to determine the value of sustainable removal of excess feedstock to farmers and potential processors across the supply chain (see Figure 2.11).
>
> A preliminary study estimated counties within a 50-mile radius of Imperial, Neb. can comply with USDA erosion control guidelines for surface cover requirements and also supply 3.6 million dry tons per year of stover with the adoption of no-till farming practices. Rail service expanded the area supply to 6 million dry tons per year with a $17-per-dry-ton margin to the farmer.

A regional supply organization, in which producers pool their harvests to provide a cohesive feedstock supply, is one way to address the high input demands of future biorefineries. For example, to supply 1.5 million dry tons requires over 500,000 acres, assuming 3 dry tons per acre excess is collected. The number of growers to reach out to for collection quickly becomes a significant and costly challenge. The first 50,000-dry-ton effort to collect corn stover near Harlan, Iowa required 400 farms and more than 30 custom harvesters to collect 30,000 acres.[16]

Figure 2.11 *Participants in the Imperial Young Farmers and Ranchers sustainable collection of biomass project stand in front of a 650 dry ton pile of corn stover.*
Source: J. Hettenhaus.

For farmers, a regional supply organization can maximize utilization of collection equipment and help attract new biorefineries by ensuring a reliable supply of feedstock.

2.9   Removal Economics

The delivered cost of cellulosic feedstocks has been estimated between $18 and $50 per dry ton. The former is based on one-pass harvesting of crop residue, collected within a 15-mile radius and shipped from collection sites to a processing plant via short line rail 200 miles or less in length. The higher value is for bales delivered within a 50-mile radius. Neither includes a margin for the farmer.

Using $50 per dry ton delivered cost, the relative economics are summarized and compared for baling (Table 2.3) and one-pass harvest, bulk storage and rail transport from remote collection sites to the processing plant (Table 2.4).[17]

Selling excess stover or straw priced at $50 per dry ton delivered may net the farmer $18 to $60 per acre if baled, depending on the yield, tillage practice, nutrient value, local situation and method of harvest. With one-pass harvest and rail transport, farmer income increases to $38 to $79 per acre.

Rail transport greatly reduces transportation costs relative to trucking, allowing for a much larger collection area. One-pass harvest, in which grain and residues are collected simultaneously, also offers strong opportunities to lower cost and reduce harvest

Table 2.3   *Net farmer income per acre for delivery of baled excess stover or straw, various yields. Assumes $50 per dry ton delivered cost, 30-mile collection radius, 1.5 million acres under cultivation, 1 dry ton per acre left in the field for soil protection, 1:1 ratio, 15% moisture*

Net farmer income (per acre) for sale of excess stover or straw – custom bale and haul			
Yield (bushels per acre)	130	170	200
Biomass available for sale (dry tons per acre)	2.0	3.0	3.8
* Sale at $50/dt	$100	$150	$190
** P & K nutrient loss to field	($12)	($19)	($24)
***Reduced field operations	$10	$10	$10
Total revenue increase	$98	$141	$176
Less custom bale, $40/ac	($40)	($40)	($40)
Handle, store, $5/dt	($10)	($15)	($19)
Shrinkage. 10%	($10)	($15)	($19)
Hauling, 30 mile radius, $10/dt	($20)	($30)	($38)
Net to farmer	**$18**	**$41**	**$60**

*The National Renewable Energy Laboratory uses $30 per dry ton delivered cost to the biorefinery as its base case scenario.[1]
**The phosphorous and potassium content in straw and stover is typically 0.1 % and 1 % respectively, valued at $6.20 per dry ton.[2] The nitrogen fertilizer value is more complex, and depends on crop rotation and local conditions.
***Reduced field operations are estimated to reduce inputs $10 per acre for preparation of the seed bed.

[1]A. Aden, M. Ruth et al., *Lignocellulosic Biomass to Ethanol Process Design and Economics Utilizing Co-Current Dilute Acid Prehydrolysis and Enzymatic Hydrolysis for Corn Stover.* NREL/TP-510-32438. (Golden, Colo.: National Renewable Energy Lab, June 2002.) http://www.nrel.gov/docs/fy02osti/32438.pdf
[2]Glassner, Hettenhaus, Schechinger, 1998.

risk. Prototypes are currently under development, funded partially by USDA and DOE projects.[18]

While the economics depend on regional and local conditions, the results serve as a template for evaluating the potential benefits. For those not currently no-tilling, the benefits of converting may justify the time to learn new methods and the $50,000 to $100,000 investment in new planting equipment. At $41 per acre (net farmer income

Table 2.4   *Net farmer income per acre for delivery of unbaled excess stover or straw with one-pass harvest and short line rail transport of residue from collection sites to processing plant, various yields. Assumes $50 per dry ton delivered cost, three 15-mile collection sites, 1.5 million acres under cultivation, 1 dry ton per acre left in the field for soil protection, 1:1 ratio, 15% moisture*

Net farmer income (per acre) for sale of excess stover or straw – custom bale and haul			
Yield (bushels per acre)	130	170	200
Biomass available for sale (dry tons per acre)	2.0	3.0	3.8
Sale at $50/dt	$100	$150	$190
P & K nutrient loss to field	($12)	($19)	($24)
Reduced field operations	$10	$10	$10
Total revenue increase	$98	$141	$176
One-pass harvest, $18/ac	($18)	($18)	($18)
Transport from field to collection site, $6/dt	($12)	($18)	($23)
Handle, store, $6/dt	($12)	($18)	($23)
Shrinkage. 3%	($3)	($5)	($6)
Rail from collection site, $7/dt	($14)	($21)	($27)
Net to farmer	**$38**	**$61**	**$79**

expected with moderate yield and custom bale and haul – Table 2.3), a 1000-acre farm could expect to recover the additional investment in as little as two years.

New markets that commoditize the environmental benefits of no-till farming could provide even greater incentive to convert to no-till cropping with one-pass harvest. These markets are developing quickly in anticipation of greenhouse gas regulation in the United States.

2.10   Climate Change Mitigation

In addition to economic benefits for farmers, sustainable production and collection of agricultural residues have the potential to deliver substantial benefits for the environment, including reduced runoff of soil and fertilizers. But perhaps the greatest environmental benefits may be to the global climate through reduced emissions of fossil carbon and enhanced sequestration of soil carbon.

The removal of crop residues by its nature reduces the amount of carbon returning to the soil, reducing the rate at which carbon is removed from the atmosphere and stored in the ground. However, studies suggest that this marginal reduction in sequestration is considerably outweighed by the reduction in fossil carbon emissions gained by the substitution of biomass for fossil-based feedstocks and by increased carbon sequestration and reduced field operations resulting from the necessary transition to no-till harvest.[19]

The potential of corn stover for greenhouse gas (GHG) mitigation is summarized in Table 2.5. Figures are for 80 million dry tons processed per year, which represents 30 % of the current annual stover production in the United States. Thirty % is a conservative estimate of the average fraction of stover that could be removed with no-till cropping, enough to produce at least 5 billion gallons of ethanol at current conversion rates.

There is a wide range in potential GHG emissions offsets, dependent on weather, soil characteristics, type of application, agronomic practices and other factors, but the potential for mitigation is substantial.

Table 2.5 *Mitigation of greenhouse gases by substitution of ethanol from corn stover for gasoline, for conversion of 30 % of available US corn stover feedstock (80 million dry tons) to ethanol*

	GHG mitigation. Conversion of 30% of available US corn stover to ethanol
Source	GHG reduction [MMT CO_2 eq]
Fossil fuel offset	50–70
Soil carbon increase[3,4]	30–50
Nitrogen fertilizer reduction	0–10
Reduced field operations[5]	10–20
Total	**90–150**

[3] Rattan Lal, John M. Kimble, Ronald F. Follett and C. Vernon Cole, *The Potential of US Cropland to Sequester Carbon and Mitigate the Greenhouse Effect* (Boca Raton, CRC Press, 1998).
[4] C.A. Cambardella and W.J. Gale, 'Carbon Dynamics of Surface Residue- and Root-derived Organic Matter under Simulated No-till.' *Soil Sci. Soc. of Amer. J.*, **64**: 190–5 (2000). D.C. Reicosky, et al., 'Soil Organic Matter Changes Resulting from Tillage and Biomass Production.' *Journal of Soil and Water Conservation*, **50**(3): 253 (May 1995).
[5] J.S. Kern and M.G. Johnson, 'Conservation Tillage Impacts on Soil and Atmospheric Carbon Levels.' *Soil Sci. Soc. Am. J.* **57**: 200–10 (1993).

The fossil fuel offset estimate, 50 to 70 million metric tons of carbon dioxide equivalent (MMTCO$_2$), is based on E85 fuel, a blend of 85 % ethanol with 15 % gasoline. The effective benefit is 0.6 to 0.9 MMTCO$_2$ mitigated per ton of corn stover processed. E85 from corn stover is estimated to reduce greenhouse gases 64 % compared to gasoline.[20]

Changing to no-till cropping can further mitigate GHG emissions by 40 to 80 million metric tons of CO$_2$ equivalent annually by increasing carbon in the soil, reducing nitrogen fertilizer needs and reducing the intensity of field operations. Collecting excess stover only from no-till fields is recommended. Tilling causes loss of soil carbon. If too much stover is removed or a cover crop is not planted, soil carbon can be depleted.

Soil carbon is thought to be more strongly impacted by below-ground residues (i.e. roots) than above-ground residue. Studies at the National Soil Tilth Laboratory show 80 % or more of the surface material is lost as CO$_2$ within months and three times the amount of soil organic matter (SOM) comes from roots compared to surface material.[21] With biorefineries, the excess surface residues are converted into fuels that power vehicles before entering the atmosphere as CO$_2$. Moving to no-till avoids the loss of SOM from plowing.

Including cover crops in the rotation helps ensure that soil quality is maintained and most likely increased before reaching a new equilibrium in 30–50 years. Cover crops can build soil carbon, reduce erosion, help control weeds and may reduce chemical inputs by controlling weeds and retaining nitrogen in the root system over the winter. However, cover crops also require a higher level of management. Selecting the appropriate cover crop to fit in the rotation can reduce cost, possibly add a third cash crop and build soil quality, especially organic material, improving yields over time. However, if not well managed, cover crops can reduce yields.[22]

Reduced nitrogen (N) fertilizer use is also possible with no-till cropping, depending on crop rotation. Soil microbes desire a 10:1 ratio of carbon to nitrogen for digesting residue. Since the carbon to nitrogen ratio of straw and stover varies between 40:1 and 70:1, nitrogen fertilizer addition equivalent to 1 % of residue is typically recommended to avoid denitrification of the next crop. When residues are removed, a more ideal ratio is maintained naturally.

For 150 bushels per acre corn yield, 70 pounds of nitrogen fertilizer may be avoided per acre if no residue is plowed under. In addition to cost savings, environmental benefits of reduced nitrogen fertilizer use include reduced run-off to streams and groundwater and reduced emissions of nitrous oxide (N$_2$O), a potent greenhouse gas. Nitrous oxide emissions range from 0.2 to 3.5 pounds of N$_2$O per 100 pounds of fertilizer applied.[23]

Since N$_2$O has 310 times the heat absorbance of CO$_2$, the resulting GHG offset is 0.5 to 9.9 metric tons CO$_2$ equivalent per ton of nitrogen fertilizer application. For 30 % of the corn stover used as feedstock, the nitrogen fertilizer reduction is 800,000 tons, for a GHG gas reduction between 1 and 10 MMTCO$_2$.

Recent work suggests that for all corn cultivation systems across the Corn Belt, nitrous oxide generated by soil microbes may be the dominant greenhouse gas emission. Cover crops cut nitrous oxide emissions by up to a factor of 10. Combining cover crops with residue removal further reduced emissions.[24]

Combined, these greenhouse gas benefits would more than offset the net growth in US emissions from all sectors of the economy in 2004.[25]

Global markets established to trade greenhouse gas emissions are now beginning to recognize no-till cropping as a legitimate tool for reducing atmospheric greenhouse gas concentrations. The Chicago Climate Exchange (CCX), a voluntary greenhouse gas trading market, allows farmers to sell the climate benefits of their cropping practice to industries wishing to offset their emissions. The Exchange offers farmers a conservative 0.5 metric ton per acre 'exchange soil offset' credit for no-till operations, which translates to roughly $1 per acre at current CCX credit prices of $1.65 to $2.00 per metric ton CO_2. The Iowa Farm Bureau and other regional growers associations have organized farmers to collectively sell their credits on the CCX market.

In European markets, where mandatory limits on greenhouse gas emissions exist, carbon credit prices have ranged between $10 and $30 per metric ton CO_2, suggesting that if mandatory greenhouse gas emissions limits are established in the United States, benefits to farmers of no-till adoption could exceed $10 per acre, further driving the transition to no-till.

2.11   Pretreatment

In addition to the challenges of sustainable production, collection and transport, cellulosic biomass presents a unique problem in that, unlike traditional starch feedstocks, it is not readily hydrolyzed into usable sugars. Cellulosic biomass must first be 'pre-treated' to make the sugars available to chemical or biological hydrolysis.

Due to nature's barriers, cellulose in biomass feedstock is 'recalcitrant' to degrade, ensheathed in protective layers of lignin and hemicellulose (see Figure 2.12).

The cell walls are intermeshed with carbohydrate and lignin polymers and other minor constituents. The major components are cellulose, hemicellulose, and lignin. These polymers have different properties, reacting in different ways to thermal, chemical, and biological processing.

Historically thermal and chemical processes overcame this recalcitrance by sheer energy input and severe chemical treatment. Typical processes included gasification, pyrolysis, Fischer-Tropsch and concentrated sulfuric acid for conversion to fuels and chemicals.[26-33] These processes developed when supplies of fossil fuels were disrupted during World Wars I and II or fuel imports embargoed under UN sanctions. Mechanical preprocessing is simple – pelletizing for thermoprocessing or commutation for concentrated acid treatment. For thermo processing, mechanical compaction increases energy density and eases bulk handling. Particle size reduction-mechanical grinding increases the surface area for improved acid processing.

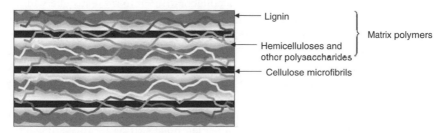

Figure 2.12   *Lignocellulosic biomass cell wall.*
Source: National Renewable Energy Laboratory.

Table 2.6 *Pretreatment process technologies*

Type	No catalyst	Acid hydrolysis	Base catalyzed	Other
Process Technology	Steam Explosion Hot Water Hot Water-pH neutral	Nitric Sulfur Dioxide Dilute Sulfuric Carbonic	AFEX Ammonia Lime	Solvent Wet Oxidation Enzymatic/Microbial Combinations

No catalyst, only steam or hot water, results in hemicellulose hydrolysis to sugars. Limited lignin is solubilized.

Acid hydrolysis (sulfuric, nitric, phosphoric, and carbonic) has similar results to steam explosion or hot water. The acid is dilute, 0.2% w/v or less. A faster hydrolysis reaction rate of hemicellulose to sugars occurs with similar conversion results of steam explosion or hot water. Limited lignin is solubilized.

Base catalyzed processes (ammonia, lime, sodium hydroxide) solubilize significant lignin but have little effect on the hemicellulose.

Others
- Wet oxidation (water, oxygen, mild alkali or acid, elevated temperature and pressure) combines several of the above to liquefy hemicellulose and lignin simultaneously.
- Organic solvents, aka Organosolv (ethanol, acetone, etc.) solubilizes lignin. While more expensive, the lignin is more active chemically than other pretreatments, offering a potentially higher value co-product. Adding a dilute acid will also remove the hemicellulose.
- Enzymatic/microbial processes reduce side reactions, increasing yield and reducing compounds that may inhibit downstream processes.

With advances in biotechnology, a global effort is underway to develop economic biological processes that break down the cellulosic components in biomass and convert them into sugars. The sugars are the platform for further biological and chemical conversion into fuels and chemicals. Lignin, mostly connected phenol groups, is high in energy, similar to Powder River Basin Coal. Commercialization plans include burning it for process energy and generating electricity. The phenol-based compounds can also meet a myriad of market needs from extending the life of asphalt paving to treating diarrhea in calves.

Biological processes occur under mild conditions when compared to thermal and chemical processes. These conditions are insufficient to expose the cellulose for effective degradation to sugars. To improve biologic process performance there are many pretreatment options under investigation. Table 2.6 summarizes the types of pretreatment process technologies that appear to be closest to commercialization.

Generally, it is difficult to compare the performance and economics of these various approaches due to difference in feedstocks tested, chemical analysis methods, and data reporting methodologies. The exception is a collaborative group of researchers funded by USDA, DOE and industry. The group, calling itself the Biomass Refining Consortium for Applied Fundamentals and Innovation (CAFI), is pursuing coordinated development of the leading biomass pretreatment technologies.[34–36] Table 2.7 tabulates the participating institutions and their processes.

A series of papers compares the features using the same corn stover feedstock, enzymes and common procedures.[37–41] Related papers are included.[42–49] The coordinated development provides a relative economic comparison of processing costs.[42] The production cost of ethanol – based on the minimum ethanol selling price (MESP) – for various pretreatment technologies is given in Figure 2.13. Minimum Ethanol Selling Price is defined as the ethanol sales price required for a zero net present value for the project when the cash flows are discounted at 10 % real-after tax.

Table 2.7 *Biomass refining Consortium for Applied Fundamentals and Innovation (CAFI) collaborative pretreatment investigators*

Institution	Process
Auburn University (Y.Y. Lee)	Ammonia Recycle Percolation (ARP)
University of British Columbia (Jack Saddler)	SO$_2$ steam explosion
University of California, Riverside (Charles Wyman)	Dilute acid treatment
Michigan State University (Bruce Dale)	Ammonia Fiber Explosion (AFEX)
Purdue University (Michael Ladisch, Nathan Mosier)	Hot water, controlled pH
Texas A&M University (Mark Holtzapple)	Lime treatment
National Renewable Energy Laboratory (NREL) (Rick Elander, Tim Eggeman)	Feed stock supply and characterization/ economic analysis

The comparison shows there is no clear choice based on operating cost. The cost range is plus or minus $0.15 per gallon, within 10 % of the average cost of $1.55 per gallon. The design is largely conceptual with considerable uncertainty in process performance. The sugar platform is based on enzyme hydrolysis of the cellulose to glucose, followed by the fermentation of the glucose to fuels and chemicals.[34]

Looking beyond the pretreatment system, there are gaps that influence the selection of 'best' pretreatment to fit the local situation. These include the following:

- feedstock supply risk
- feedstock logistics
- farmers' participation in cellulosic feedstock value chain
- co-products and added value.

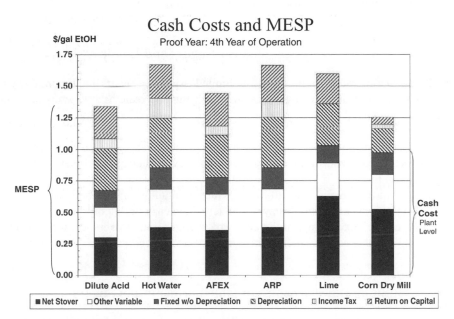

Figure 2.13 *Cash cost and minimum ethanol selling price (MESP) per gallon for various pretreatment technologies.*

Source: Biomass Refining Consortium for Applied Fundamentals and Innovation (CAFI).

2.12   Farmer in Value Chain

Farmer-owned dry mill ethanol plants are a shining example of how farmers can prosper from down stream 'value added' processing. In the areas surrounding these plants the farmers have enjoyed a stronger market for their corn since the plants began operation. In addition, operating margins from their plants have increased due to higher gasoline pricing. The farmer-investors and others in the community are benefiting from the increased dividends and higher value of their investment. The local economy has also benefited from the jump in local expenditures to improve and expand plant operations.

Can processing cellulosic feedstocks produce similar benefits to the farmer and local communities? If not, will expectations for renewable biomass feedstocks be realized?

Most cellulosic business models envision a large cellulosic biorefinery processing 700–2000 dry tons per day of feedstock supplied from fields of local farmers. The biorefinery plant investment is estimated to be $200 million to $400 million. How many farmers are going to be able to own one of these cellulosic plants to convert the bulky material to ethanol? Conversely, can the plant insure a reliable and economic feedstock supply over the life of the plant without farmers participating in the value added to the supply chain?

However, if the farmers collect and convert the biomass to liquids using ammonia explosion, hot water, dilute acid and a base resulting in fermentation sugars, soluble lignin and other liquids (lime treatment), the equity required is estimated to be $15 million or less, well within the range for farmer ownership.[50]

Preprocessing in the field can take on several operations. First, biomass is collected and stored in large, wet piles similar to sugar cane bagasse during the harvest.[51–54] Adjacent to the biomass storage, a modular pre processing system is easiest to envision. Modules can be added to expand the quantity and degree of biomass processed.

2.13   The Start: Preprocessing Pentose Sugars and Lignin

Initially, the liquid volume could simply be limited to pentose sugars and lignin. Easiest is removing the hemicellulose with acid-enzyme hydrolysis, producing a 30 % dry substance pentose sugar stream that can be fed directly to fermentors in the biorefinery down the road or at the other end of the pipe.

Next is making the lignin soluble. The soluble lignin is transported, precipitated, filtered and used as a process energy source by the biorefinery, replacing fossil fuels or the more bulky crop residues.

Hydrolysis of the remaining cellulose to a glucose solution can be readily accomplished at the collection center by adding enzymes or activating enzymes incorporated in the plant. Depending on the enzyme hydrolysis economics, transpiration of the cellulosic solids – less than 2/3 the original feedstock solids – may be a viable option.

2.14   Continuing Downstream: Fungible Fermentation Sugars

As the demand for biomass feedstock market grows and farmers collect more biomass, additional modules can be added to expand capacity and processing. Continuing to move downstream in adding value (upstream from the biorefinery), hydrolyzing the cellulose completes the cellulosic conversion to sugars at the storage site. Depending on the local situation, the use of dilute fermentation sugars may be limited. Concentrating the sugars from 30 % to 80 % solids would likely make them fungible, opening up a broader market to supply other biorefineries. Concentrated fermentation sugars reduce feedstock risk, removing a major concern when considering debt financing.

2.15   Looking Upstream

Farmers can work with crop science companies – such as Monsanto, Pioneer and Syngenta for corn and soybeans or Ceres and Mendel Biotechnology for perennial grasses – to obtain new cropping systems with traits that are desirable for increasing the margins from the crops and achieve sustainable production.

Companies looking at plants as sources for chemicals and plastics, like Metabolix, ADM, National Starch and Allylix have a potential partner for growing the desired crops, recovering the expressed 'value added' material and processing the remainder.

2.16   Logistics

Transporting and delivering liquids to biorefineries like dry mill ethanol plants eases the logistic problem. The liquids can be delivered via truck or rail, but pipelines to transport the liquids – similar to the pattern used for oil wells to collect fossil fuels – avoid traffic disruption and have an economic advantage in most places.

Expanding the fermentation and downstream processing operations of existing ethanol plants permits capacity growth without going beyond their present corn buying area and increases their competitive position – benefiting from economies of scale.

A gasifier using some of the biomass as a feedstock can provide the process energy source. As markets develop for lignin products beyond fuels, selecting plants with the desired lignin properties and recovering lignin as a co-product can provide additional value added.

Providing fermentation sugars and lignin feedstocks in lieu of crop residues positions the farmer for benefiting from Greenhouse gas offset credits when carbon trading becomes a standard way of doing business. The economic value is expected to be 5–10 times the $2–$4 per ton of CO_2 now being offered to US farmers for carbons offsets.

The engineering and biotech tools are available to make this happen. When the economics work, we can expect the farmers will be a huge player in the future – the carbohydrate version of OPEC, as they control the feedstock.

2.17   Conclusions

Higher and more unstable prices of petroleum and natural gas, along with the desire to produce more energy domestically, are making agricultural feedstocks a more attractive alternative for the transportation fuel, chemicals, and plastics industries. With ongoing advances in both agricultural and industrial biotechnology, the economics are improving for these industries to switch from petrochemical to agricultural feedstocks. Governmental and international policies to reduce GHG emissions will add further economic incentives for biomass utilization as they develop.

Issues surrounding harvest, storage, transportation and pre-processing of feedstock supply need to be given significantly more attention if biomass is to serve as a sustainable platform for this industrial shift. Providing biomass feedstocks in sufficient and steady quantities to a biorefinery will require sustainable production and harvest practices as well as improved methods for pre-processing of biomass and delivery of biomass to the processor from the field.

In the near to mid term, crop residues are most likely to be the feedstock of choice for biorefineries. Improved agronomic systems – such as no-till cropping – with more crop residue removal can be implemented while also maintaining soil quality. The development of biorefineries, providing a market for crop residues, will provide economic incentives to farmers to adopt no-till cropping methods.

There are significant environmental benefits that can be gained from the switch to agricultural from petroleum feedstocks for the transportation and chemical industries. Farmers who supply agricultural feedstocks to these industries could also benefit from a carbon credit system by switching to no-till cropping methods.

Coordinated governmental agriculture and energy policies are needed to encourage the growth of the biorefinery industry and facilitate the sustainable production of agricultural feedstock supplies.

2.18   Policy Recommendations

As a result of research for this report and participation in several colloquies sponsored by federal agencies, I have developed a menu of recommendations that may help facilitate development of the infrastructure necessary for sustainable production and collection of cellulosic agricultural feedstocks, and achieve the DOE goal of 30 % displacement of petroleum with renewable biobased feedstocks by 2030. Congress should consider implementing the following policy measures:

- Fund research and development and provide assistance for the purchase of one-pass harvesting equipment and other new harvesting equipment for collection of cellulosic agricultural feedstocks.
- Develop and make available simple-to-use soil carbon computer models to allow individual farmers to compute how much crop residue can be collected without degrading soil quality.
- Provide assistance to farmers to encourage the transition to no-till cropping for biomass production.
- Provide incentives for the development and expansion of short line and regional rail networks for transport of cellulosic feedstocks.
- Fund regional demonstration projects to streamline the collection, transport and storage of cellulosic feedstocks.

- Develop a system to monetize greenhouse gas credits generated by production of ethanol and other products from agricultural feedstocks.

Cellulosic biomass from agricultural residues represents a highly promising new source of feedstock material for the production of ethanol, renewable chemicals and a range of commercial products. Residues from existing crops can greatly expand current production. American farmers are poised to deliver.

References

1. Energy Information Administration (year?) www.eia.doe.gov/emeu/steo/pub/contents.html.
2. US Department of Energy (2004) 'Top Value Added Chemicals from Biomass, Volume I – Results of Screening for Potential Candidates from Sugars and Synthesis Gas.' (Washington, D.C., DOE, August 2004.) http://www1.eere.energy.gov/biomass/pdfs/35523.pdf.
3. National Corn Growers Association (2006) 'How Much Ethanol Can Come from Corn?' (Washington, D.C., NCGA) http://www.ncga.com/ethanol/pdfs/2006/HowMuchEthanolCan%20ComeFromCorn.v.2.pdf.
4. Robert D. Perlack, Lynn L. Wright et al. (2005) 'Biomass as Feedstock for a Bioenergy and Bioproducts Industry: The Technical Feasibility of a Billion-Ton Annual Supply.' ORNL/TM-2005/66. (Oak Ridge, Tenn., ORNL, April 2005) http://feedstockreview.ornl.gov/pdf/billion_ton_vision.pdf.
5. Multi Year Program Plan, 2007–2012 (2005) (Office of Biomass Programs, EERE, DOE, Aug. 31, 2005.) http://www1.eere.energy.gov/biomass/pdfs/mypp.pdf.
6. US Department of Energy (2006) 'Breaking the Biological Barriers to Cellulosic Ethanol: A Joint Research Agenda.' DOE/SC-0095. (Washington, D.C., US Department of Energy, Office of Science and Office of Energy Efficiency and Renewable Energy, June 2006.) http://doegenomestolife.org/biofuels/b2bworkshop.shtml.
7. Hettenhaus, James (2004), Biomass Feedstock Supply: As secure as the pipeline to a Naphtha Cracker? Presentation at World Congress on Industrial Biotechnology and Bioprocessing, Orlando, FL, April 21–23, 2004.
8. James Hettenhaus, Robert Wooley, John Ashworth (2002) 'Sugar Platform Colloquies,' Subcontractor Report, NREL/SR-510-31970. (Golden, Colo., National Renewable Energy Laboratory, May 2002.) http://www1.eere.energy.gov/biomass/pdfs/sugar_platform.pdf.
9. Hettenhaus, James, Reed Hoskinson and William West (2003) Feedstock Roadmap Colloquies Report: Feedstock Harvesting and Supply Logistics, Research and Development Roadmap. INEEL PO No. 00018408. (Idaho Falls, Idaho: Idaho National Engineering & Environmental Laboratory, Nov. 2003.) http://www.ceassist.com/pdf/feedstock_roadmap_colloquy.pdf#search=%22feedstock%20colloquies%20roadmap'%22.
10. Sustainable Agriculture Network (2001) Managing Cover Crops Profitably, 2nd Ed. (Beltsville, Md., Sustainable Agriculture Network, National Agricultural Library) http://www.sare.org/publications/covercrops/covercrops.pdf.
11. Mann, L., Tolbert, V. and Cushman, J. (2002) 'Potential environmental effects of corn (Zea mays L.) stover removal with emphasis on soil organic matter and erosion.' *Agriculture, Ecosystems and Environment* **89**: 149–66.
12. Conservation Technology Information Center (2006) 'National Crop Residue Management Survey: Conservation Tillage Data.' (W. Lafayette, Ind., CTIC) http://www2.ctic.purdue.edu/CTIC/CRM.html.
13. USDA, Energy and Agriculture (2007) Farm Bill Theme Paper. (Washington, D.C., August 2006). www.usda.gov/documents/Farmbill07energy.pdf.

14. Cavener, Lorraine (2004) 'Stalking Future Feedstocks' Ethanol Producer Magazine. (Grand Forks, N.D., March.) http://www.ethanolproducer.com/article.jsp?article_id=1155& q=stalking%20future%20feedstocks&category_id=29.

15. Worldwatch Institute (2006) 'Biofuels for Transportation: Global Potential and Implications for Sustainable Agriculture and Energy in the 21st Century.' (Washington, D.C.) http://www. worldwatch.org/taxonomy/term/445.

16. Glassner, D., J. Hettenhaus, T. Schechinger (1998) 'Corn Stover Collection Project,' BioEnergy '98 – Expanding Bioenergy Partnerships: Proceedings, Volume 2, Madison, WI, pp 1100–10.

17. J.E. Atchison and J.R. Hettenhaus (2003) Innovative Methods for Corn Stover Collecting, Handling, Storing and Transporting. NREL/SR-510-33893. (Golden, Colo.: National Renewable Energy Lab, March) http://www.nrel.gov/docs/fy04osti/33893.pdf.

18. Quick, Graeme (2003) Single-Pass Corn and Stover Harvesters: Development and Performance. Proceedings of the International Conference on Crop Harvesting and Processing, 9–11 February 2003 (Louisville, Kentucky USA) 701P1103e.

19. Sheehan, John, Andy Aden, Keith Paustian, Kendrick Killian, John Brenner et al. (2003) Energy and environmental aspects of using corn stover for fuel ethanol, *J. Ind. Ecology* 7(3–4).

20. Levelton Engineering Ltd. and (S&T)2 Consultants Inc. (2000) 'Assessment of Net Emissions of Greenhouse Gases From Ethanol-Blended Gasoline in Canada: Lignocellulosic Feedstocks,' Report to Agriculture and Agri-Food Canada. (Ottawa, Ontario, Agriculture and Agri-Food Canada, January 2000.) http://www.tc.gc.ca/programs/Environment/climatechange/docs/biomass/JanFinalBiomassReport.htm.

21. C.A. Cambardella and W.J. Gale (2000) 'Carbon dynamics of surface residue– and root-derived organic matter under simulated no-till,' *Soil Sci. Soc. of Amer. J.*, **64**: 190–5.

22. D.C. Reicosky et al. (1995) 'Soil organic matter changes resulting from tillage and biomass production,' *Journal of Soil and Water Conservation* **50**(3): 253 (May).

23. Council for Agricultural Science and Technology (1992) 'Preparing U.S. Agriculture for Global Climate Change,' Task Force Report 119. (Ames, Iowa: Council for Agricultural Science and Technology).

24. Bruce Dale, Michigan Statue University, unpublished.

25. US Environmental Protection Agency (2006) 'Inventory of U.S. Greenhouse Gas Emissions and Sinks: 1990–2004,' (US EPA, Washington, D.C., April). http://yosemite.epa.gov/oar/globalwarming.nsf/content/ResourceCenterPublicationsGHGEmissionsUSEmissionsInventory2006.html

26. Bain, R.L., Amos, W.P., Downing, M., Perlack, R.L. (2003) Biopower Technical. Assessment: State of the Industry and the Technology. 277 pp.; NREL Report No.TP-510-33123.

27. Bridgewater, A.V. (2003) A Guide to Fast Pyrolysis of Biomass for Fuels and Chemicals, PyNe Guide 1, www.pyne.co.uk

28. Higman, C. and M. van der Burgt (2003) *Gasification*, Elsevier Science (USA), ISBN 0-7506-7707-4.

29. Probstein, R.F. and R.E. Hicks (1982) *Synthetic Fuels*, McGraw-Hill, Inc., ISBN 0-07-050908-5.

30. Knoef, H.A.M. (ed.) (2005) *Handbook Biomass Gasification*, BTG Biomass Technology Group, Enschede, The Netherlands, ISBN 90-810068-1-9.

31. Mills, G. (1993). 'Status and Future Opportunities for Conversion of Synthesis Gas to Liquid Energy Fuels: Final Report,' NREL,TP-421-5150, National Renewable Energy Laboratory, Golden, CO, May.

32. Spath, P.L., and D.C. Dayton (2003) 'Preliminary Screening – Technical and Economic Assessment of Synthesis Gas to Fuels and Chemicals with Emphasis on the Potential for Biomass-Derived Syngas, NREL Report TP51034929, National Renewable Energy Laboratory, Golden, CO September.

33. Wright, J.D., d'Agincourt, C.G. (1984) 'Evaluation of Sulfuric Acid Hydrolysis Processes for Alcohol Fuel Production.' *Biotechnology and Bioengineering Symposium,* No 14, John Wiley & Sons, Inc., New York, 1984, pp. 105–23.

34. Wyman, C.E., Dale, B.E., Elander, R.T., Holtzapple, M., Ladisch, M.R., Lee, Y.Y. (2005), Coordinated development of leading biomass pretreatment technologies, *Bioresource Technology* **96**(18) (Dec.): 1959–66.

35. Mosier, N., Wyman, C., Dale, B., Elander, R., Lee, Y.Y. et al. (2005) Features of promising technologies for pretreatment of lignocellulosic biomass, *Bioresource Technology* **96**(6) (Apr.): 673–86.

36. Wyman, C.E., Dale, B.E., Elander, R.T., Holtzapple, M., Ladisch, M.R., Lee, Y.Y. (2005) Comparative sugar recovery data from laboratory scale application of leading pretreatment technologies to corn stover, *Bioresource Technology* **96**(18) (Dec.): 2026–32.

37. Liu, C., Wyman, C.E. (2005) Partial flow of compressed-hot water through corn stover to enhance hemicellulose sugar recovery and enzymatic. *Bioresource Technology* **96**(18) (Dec.): 1978–85.

38. Mosier, N., Hendrickson, R., Ho, N., Sedlak, M., Ladisch, M.R. (2005) Optimization of pH controlled liquid hot water pretreatment of corn stover, *Bioresource Technology* **96**(18) (Dec.): 1986–93.

39. Kim, S., Holtzapple, M.T. (2005) Lime pretreatment and enzymatic hydrolysis of corn stover, *Bioresource Technology* **96**(18) (Dec.): 1994–2006.

40. Teymouri, F., Laureano-Perez, L., Alizadeh, H., Dale, B.E. (2005) Optimization of the ammonia fiber explosion (AFEX) treatment parameters for enzymatic hydrolysis of corn stover, *Bioresource Technology* **96**(18) (Dec.): 2014–18.

41. Kim, T.H., Lee, Y.Y. (2005) Pretreatment and fractionation of corn stover by ammonia recycle percolation process, *Bioresource Technology* **96**(18) (Dec.): 2007–13.

42. Eggeman, T., Elander, R.T. (2005) Process and economic analysis of pretreatment technologies, *Bioresource Technology* **96**(18) (Dec.): 2019–25.

43. Kim, T. H., Y.Y. Lee, C.S. Sunwoo, and J. S. Kim (2006) Pretreatment of corn stover by low-liquid ammonia recycle percolation process, *Applied Biochemistry and Biotechnology* **133**.

44. Kim, T.H., Y.Y. Lee (2005) Pretreatment of corn stover by soaking in aqueous ammonia, *Applied Biochemistry and Biotechnology* 121–4.

45. Peter van Walsum, G., Shi, H. (2004) Carbonic acid enhancement of hydrolysis in aqueous pretreatment of corn stover, *Bioresource Technology* **93**(3) (July): 217–26.

46. Zimbardi, F., Viola, E., Nanna, F., Larocca, E., Cardinale, M., Barisano, D. (2007) Acid impregnation and steam explosion of corn stover in batch processes, *Industrial Crops & Products* **26**(2) (Aug.): 195–206.

47. Bobleter, O., Bonn, G., Prutsch, W. (1991) Steam explosion-hydrothermolysis-organosolv. A comparison. In: Focher B, Marzetti A, Crescenzi V. (eds), *Steam Explosion Techniques.* Gordon & Breach, Philadelphia, pp. 59-82.

48. Varga, E., Reczey, K., Zacchi, G. (2004) Optimization of steam pretreatment of corn stover to enhance enzymatic digestibility. *Appl. Biochem. Biotech.* 113–16, 509–23.

49. Chang, V.S., Nagwani, M., Kim, C.H., Holtzapple, M.T. (2001) Oxidative lime pretreatment of high-lignin biomass, *Appl. Biochemistry and Biotechnology* **94**: 1–28.

50. Carolan, J.E., S.V. Joshi, B.E. Dale (2007) Explorations in biofuels economics, policy, and history: technical and financial feasibility analysis of distributed bioprocessing using regional biomass pre-processing centers, *Journal of Agricultural & Food Industrial Organization* **5**(10).

51. Salaber, J., Maza (1971) Ritter Biological Treatment Process for Bagasse Bulk Storage. TAPPI Non-wood Plant Fiber Pulping Progress Report, No 2, October.

52. Moebius, J. (1966) The Storage and Preservation of Bagasse in Bulk Form, Without Baling. Pulp and Paper Development in Africa and the Near East, United Nations, N.Y. Volume II.

53. Atchison, J. (1972) Review of Progress with Bagasse for Use In Industry (A review of progress in purchasing, handling, storage and preservation of bagasse) *J.E. Proc. Intern. Soc. Sugar Cane Technologists* **14**: 1202–17 (1971) Franklin Press, Baton Rouge, LA.
54. Hettenhaus, J.R. (2006) Biomass Commercialization and Agriculture Residue Collection, Chapter 3.1, *Biorefineries, Biobased Industrial Processes and Products* Ed: B. Kamm, P. Gruber, M. Kamm, Wiley-VCH Verlag GmbH & Co.

3

Bio-Ethanol Development in the USA

Brent Erickson

Executive Vice President, Industrial and Environmental Section, Biotechnology Industry Organization, Washington, USA

Matthew T. Carr*

Policy Director, Industrial and Environmental Section, Biotechnology Industry Organization, Washington, USA

3.1 Introduction

From the First Continental Congress, which met over fine spirits at Philadelphia's City Tavern in 1774, the history of the United States of America is intersected by ethanol. But not until the early 20th century, when the US Congress removed a $2 per gallon ethanol tax first established to fund the Civil War, did bio-ethanol began to develop as a commercially viable liquid transportation fuel.

One of bio-ethanol's first US champions was Ford Motor Company founder Henry Ford. He designed the first Model T automobile to run on 100 % ethanol. 'The fuel of the future is going to come from fruit like that sumac out by the road, or from apples, weeds, sawdust – almost anything. There is fuel in every bit of vegetable matter that can be fermented,' Ford told the *New York Times* in 1925.[1]

---

*The authors wish to thank Joanne Hawana for her assistance in researching this chapter.

By the 1930s, fuel ethanol had gained a significant market in the US Midwest. At one point over 2000 fuel stations sold 'gasohol' – gasoline blended with up to 12 % ethanol. But advances in the production of gasoline soon priced ethanol out of the transportation fuel marketplace, and between 1950 and the late 1970s virtually no commercial fuel ethanol was sold in the US.[2]

The oil shocks and growing environmental awareness of the 1970s changed that, and federal government policies designed to lessen US dependence on fossil fuels – particularly foreign petroleum – have grown the US ethanol industry ever since.

3.2   Federal Policy

Much of bio-ethanol's revival in the US can be traced to federal legislation that resulted from the oil crises of 1973 and 1979. Petroleum prices in the US quadrupled from 1973 to 1974, then tripled from 1978 to 1979, following supply disruptions in the Middle East, spurring a flurry of federal policy to reduce energy demand and encourage alternative energy production.[3]

But even before the Arab oil embargo, a re-examination of ethanol was underway. In early 1973, growing health and environmental concerns led the US Environmental Protection Agency to issue regulations that would eventually result in the complete elimination of lead as a gasoline additive.[4] Petroleum refiners began to look for alternatives that could provide the same level of octane enhancement as lead, and ethanol provided an attractive option.

In 1978, in an effort to reduce dependence on foreign oil and grow domestic production of alternative fuel, the US Congress exempted ethanol-blended gasoline from the federal fuel excise tax, providing what amounted to a subsidy of $0.40 per gallon for ethanol.[2] Marketing of commercial ethanol-blended fuels began the next year.

Numerous additional incentives for ethanol production were introduced in the early 1980s, including additional tax benefits, loan guarantees and purchase agreements. The Congress also imposed an import tariff, protecting US manufacturers from inexpensive Brazilian ethanol.

By 1984, there were 163 ethanol plants in the US with production capacity of nearly 600 million gallons.[2] But in 1985, petroleum prices dropped dramatically to pre-1973 levels. Even with an enhanced subsidy of 60 cents per gallon, nearly half of commercial ethanol plants went out of business.

In 1990, the US Congress moved to further reduce air pollution with a series of amendments to the federal Clean Air Act. Among these was a new requirement that oxygen be added to gasoline to reduce carbon monoxide emissions. Ethanol served this purpose nicely, but so too did methyl tertiary butyl ether (MTBE), a volatile hydrocarbon-based solvent. Most US fuel blenders opted for MTBE, since it could be produced from byproducts of the petroleum refining process. But it soon became apparent that the additive had its own environmental issues.

The US experienced a rash of ground water pollution incidents in the 1990's as old underground gasoline tanks at fueling stations, pipe connections, and other infrastructure began to leak. Because MTBE is a volatile ether, its odor and taste is detectable in drinking water even at parts per million. MTBE contamination of ground and surface water became a national environmental issue.

Oil companies had strenuously opposed any requirements for the use of ethanol fuels prior to the widespread environmental problems caused by MTBE. But the threat of MTBE contamination lawsuits and the growing list of states and local governments banning MTBE use, helped to soften opposition. Bio-ethanol production doubled over the period of 2000 to 2004 despite a reduction in the ethanol excise tax credit (now known as the Volumetric Ethanol Excise Tax Credit – VEETC) to 51 cents per gallon.[5]

In 2005, in a sweeping piece of energy legislation, the Congress declined to provide fuel blenders with protections from MTBE contamination lawsuits – effectively eliminating its use nationwide – and instituted a national ethanol mandate known as the Renewable Fuels Standard (RFS) that required a doubling of ethanol production to 7.5 billion gallons by 2012. Demand for ethanol surged.

The law, known as the Energy Policy Act of 2005 (EPACT), also established nearly $4 billion in programs and incentives to accelerate the commercialization of a new generation of ethanol from 'cellulosic' biomass. Members of Congress have embraced cellulosic ethanol – produced from agricultural residues such as corn stalks or wheat straw, or dedicated energy crops such as perennial grasses or fast-growing trees, as an opportunity to extend the economic development benefits of ethanol production to all 50 states. The proliferation of cellulosic ethanol provisions in EPACT reflects this enthusiasm – though many of the law's programs have yet to be funded (see Table 3.1).

In addition to the benefits that new energy policy legislation provided, there was increasing concern in the White House about the negative national security implications of US dependence on foreign petroleum. Immediately following passage of EPACT, President George W. Bush vigorously threw his support behind biofuels. President Bush shone a national spotlight on cellulosic ethanol in his 2006 State of the Union address as a solution to what he described as America's 'addiction to oil', further fueling enthusiasm for the technology. A growing number of individual states have also begun providing incentives to attract cellulosic ethanol production, and the newly Democratic Party-controlled Congress is expected to provide even further inducements sometime this year.

3.3   The US Ethanol Market

The recent surge of new and very supportive government policies, combined with increasing petroleum prices, has produced an American ethanol boom. Ethanol production in the US has more than tripled since 2000 (see Figure 3.1). Today, more than 150 ethanol biorefineries are in operation in the United States with a combined annual capacity of nearly 9 billion gallons. More than 50 additional facilities are now under construction or expansion with capacity to double US ethanol production to over 12 billion gallons by 2009.[6]

In 2006, ethanol producers enjoyed record profits. Market prices for ethanol exceeded $5.00 a gallon during the summer months – nearly triple 2005 levels.[7] With an average production cost of $1.45 per gallon,[8] ethanol plant developers recovered the cost of plant construction in as little as six months. With the influx of new production, market prices for ethanol have since returned closer to their historic average of $2.00 per gallon.[9]

Table 3.1   Summary of bio-ethanol provisions in the Energy Policy Act of 2005

Bio-ethanol Provisions in the Energy Policy Act of 2005 (EPACT)
The Energy Policy Act of 2005 was signed into law by President George W. Bush on August 8, 2005. The bill doubles the volume of renewable fuels in the nation's fuel supply by 2012, provides tax incentives for biofuels production and distribution, and establishes nearly $4 billion in research and demonstration programs for the development of cellulosic ethanol.

Commercialization/government procurement

• Creates a **Renewable Fuels Standard** (**RFS**) requiring fuel refiners to double the volume of biofuels in the nation's fuel supply to 7.5 billion gallons annually by 2012. Starting in 2013, a minimum of 250 million gallons a year must be cellulosic ethanol.
• Sets up a program of **loan guarantees** to reduce investment risk to private lenders for construction of cellulosic ethanol biorefineries. [First loan guarantees expected to be issued in 2007.]
• Authorizes $250 million in **production incentives** for first 1 billion gallons of cellulosic biofuels production through a 'reverse auction' process. [Yet to be funded.]
• Requires the use of renewable fuels in all government vehicles capable of their use.

Tax incentives

• Tax credit to fueling stations for 30 percent of the cost of installing alternative fuel pumps.
• 10 cent per gallon small ethanol producer credit to plants with capacity under 60 million gallons.

R&D

• Funds bioenergy/biorefinery research, development and demonstration up to $2.9 billion over 10 years through three separate programs:
  • **DOE Bioenergy Program** – $738 million authorized over 3 years. Includes $375 million for Integrated Biorefinery Demonstration. [Six biorefinery demonstration grants issued in early 2007.]
  • **USDA Biomass R&D Program** – $2 Billion for production, collection, transport and pre-processing of cellulosic crops. [Yet to be funded.]
  • **DOE Office of Science** – $196 million for development of cellulase enzyme cocktails and advanced fermentation organisms for cellulosic feedstocks. [DOE currently reviewing bids to establish two National Bioenergy Research Centers.]

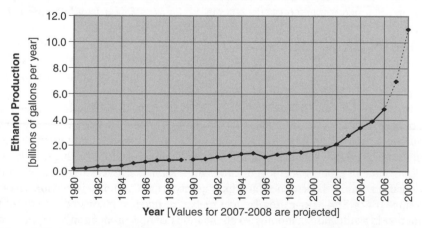

Figure 3.1   Historical US ethanol production.
Source: Ref.[5].

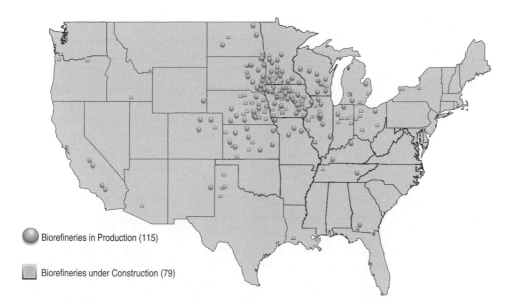

Figure 3.2   *Map of US ethanol biorefinery locations.*

Source: Renewable Fuels Association, http://www.ethanolrfa.org/objects/documents/plantmap_040307.pdf

The vast majority of current ethanol facilities (over 85 %) are situated in the Midwest United States of Illinois, Indiana, Iowa, Kansas, Michigan, Minnesota, Missouri, Nebraska, North Dakota, Ohio, South Dakota, and Wisconsin (see Figure 3.2). The Midwest is the heartland of US agricultural production. With its rich soils and warm, humid summers, the Midwest is home to the highest concentration of corn production in the US, and in the world of US bio-ethanol corn is king.

Starch from corn kernels is the feedstock for 98 % of US ethanol production. An estimated 1.6 billion bushels, or 14.4 % of the total corn harvest for 2005, was used for ethanol production during the period of September 2005 to August 2006. Ethanol production is expected to require 20 % of total corn production for the period of September 2006 through August 2007, with as much as one-third of the total corn grain harvest going to ethanol production by the end of the decade.[10] Grain sorghum (milo), barley, wheat and food/beverage residues account for the remaining 2 % of current US ethanol feedstocks.[6]

The National Corn Growers Association projects that with continued advances in agricultural biotechnology as many as 6 billion bushels of US corn grain could be available for ethanol production by 2015 while continuing to satisfy food, animal feed and export demands. That amount of corn could produce nearly 18 billion gallons of ethanol, enough to meet over 10 % of projected US gasoline demand.[11]

A growing number of non-governmental organizations have begun to express concerns about the impact such an increase in corn production would have on wildlife and the environment. Livestock producers, food processors and consumer groups have also expressed concerns that demand for corn will drive up prices for food and animal feed. Corn prices have nearly doubled since 2006.[9] but long-term effects on agricultural commodity markets remain to be seen.

Despite these concerns, a growing list of corn ethanol facilities is also emerging outside the traditional Midwest corn belt. California's central valley, which is another strong corn-producing area, now features several biorefineries to feed California's voracious appetite for fuel. Texas, a state usually synonymous with petroleum, is also experiencing an ethanol boom. Northwest Texas has only a modest corn harvest, but biorefinery developers have been lured by the abundant supply of renewable power available from the region's high-density cattle operations – cow manure is an emerging alternative to natural gas and coal as a climate-friendly non-fossil source of heat and power for ethanol production.

Other facilities, such as those under construction along the Columbia River separating Washington State and Oregon, and two planned facilities along the shores of the Great Lakes in New York State, have been sited to take advantage of existing transportation infrastructure. Transportation – both of feedstocks and finished product – is a considerable component of the cost of ethanol production. The ability to ship by barge or rail is increasingly important as the cost of gasoline in the US continues to rise.

While corn is likely to remain the dominant ethanol feedstock for the duration of this decade, high fossil energy costs are now converging with breakthroughs in biotechnology and supportive government policy to drive a new generation of ethanol production. Cellulosic ethanol, produced from lignocellulosic biomass found in agricultural residues such as corn stalks or wheat straw, or in dedicated energy crops such as switchgrass or hybrid tree varieties, has tremendous potential to greatly expand ethanol production in the United States.

A comprehensive analysis by the US Department of Energy (DOE) and the US Department of Agriculture (USDA)[12] found that, through collection of agricultural, municipal and forestry residues and a shift towards greater use of dedicated energy crops, more than 1.3 billion dry tons of biomass could be available on an annually sustainable basis for biofuels production by 2030. This volume of biomass would be enough to supply over half of current US gasoline demand of 145 billion gallons annually, even at current early-stage ethanol yields for cellulosic biomass.[13] Nearly one billion dry tons of this could be produced by American farmers.

By far the largest near-term source of available cellulosic biomass is corn stover – the stalks and leaves that remain on the field after corn kernels are harvested (see Figure 3.3). A fraction of this stover must remain on the fields to maintain soil quality and prevent erosion, but an estimated 75 million to 200 million dry tons of excess corn stover could be sustainably collected each year in the United States, enough to triple current ethanol production.[14] Corn stover grows at a very high density – yields can exceed 5 dry tons per acre, 3–5 times greater than cereal straw for example – making it an attractive source of biomass. However, corn stover also presents several challenges for collection and processing.

Unlike wheat straw, which already has an established harvesting and collection infrastructure, very little corn stover is currently collected. Potentially costly new harvesting and baling equipment will likely be needed, which could hinder farmer participation initially. Corn stover also contains as much as 50 % moisture and must remain in the field to dry before collection. Research is underway to address these concerns.

In the meantime, despite its lower yields, the presence of an existing collection infrastructure is likely to mean that cereal straw, primarily from wheat, will be one of the first cellulosic biomass crops to be collected for biofuels production. A variety of waste resources, including wood and pulp mill waste and municipal solid waste also offer promising sources of cellulosic biomass.

USDA/DOE Estimated Current Sustainable Availability of Cellulosic Biomass from Agricultural Lands

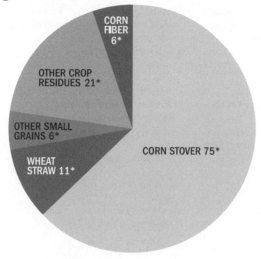

* Figures above represent millions of dry tons per year.

Figure 3.3 *Current sustainable availability of cellulosic biomass from agricultural lands.*
Source: Ref.[12].

In a major step forward, the US Department of Energy announced in early 2007 six grants totaling $385 million to construct the first commercial cellulosic ethanol biorefineries in the United States. The federal government will provide up to 40 % of the cost of biorefinery construction for each project. With a combined capacity of over 130 million gallons annually, these facilities will demonstrate a variety of processing technologies using a variety of feedstocks in six distinct geographic areas (see Table 3.2).

In May 2007, the Department also announced an additional $200 million in grants for construction of smaller pilot-scale biorefineries to develop new processes for converting other cellulosic feedstocks. The DOE is also expected to begin issuing loan guarantees authorized under EPACT to encourage private lenders to invest in cellulosic biorefinery construction.

President Bush has set a national goal of 30 % displacement of petroleum in the transportation sector by 2030, which would require approximately 60 billion gallons of bio-ethanol or other gasoline alternatives annually. With a projected upper limit of 15–18 billion gallons of ethanol from corn starch, much depends on the successful commercialization of cellulosic ethanol production.

One of the challenges for the ethanol industry is the current ethanol infrastructure in the United States. Presently, fewer than 3 % of the 230 million vehicles on US roads are capable of operating on ethanol blends higher than 10 %.[15] Thus, the current domestic market for ethanol is limited to approximately 15 billion gallons annually – 10 % of the gasoline market. Corn is capable of supplying nearly all of the ethanol up to this 'blend

Table 3.2   *2007 U.S. Department of Energy Biorefinery grant recipients*

Company	Plant location	Feedstocks	Technology	Capacity
Abengoa Bioenergy	Kansas (co-located with corn grain plant)	Corn stover, wheat straw, milo stubble, switchgrass	Enzymatic Hydrolysis	11 Mgal + power
ALICO	Florida	Yard, wood, vegetative wastes	Gasification-Fermentation	14 Mgal + power, hydrogen, ammonia
BlueFire Ethanol	California (at existing landfill)	Sorted green waste and wood waste from landfills	Acid Hydrolysis	19 Mgal
Poet	Iowa (at existing corn grain plant)	Corn fiber, cobs, and stalks	Enzymatic Hydrolysis	30 Mgal
Iogen	Idaho	Wheat straw, barley straw, corn stover, switchgrass, and rice straw	Enzymatic Hydrolysis	18 Mgal
Range Fuels	Georgia	Wood residues and wood-based energy crops	Gasification	40 Mgal ethanol + methanol

Source: US Department of Energy.

wall', leaving little surplus demand for cellulosic production. To address this barrier, the US Congress is currently considering a new ethanol mandate to require the use of up to 35 billion gallons of ethanol by 2022. This would be a significant policy driver for rapid commercialization of cellulosic ethanol in the future.

US automakers have also pledged to make half the cars they manufacture compatible with E85 (85 % ethanol; 15 % gasoline blend) by 2012, vastly expanding the market for higher-percentage ethanol blends. But such 'flexible fuel' vehicles (FFVs) will have a hard time finding anywhere to fill their tanks. Presently, fewer than 1500 filling stations across the country offer E85 fuel[16] – less than the number that offered gasohol in the 1930s. Lawmakers at federal and state levels are considering mandates and incentives to increase both the number of FFVs and the number of filling stations offering E85 fuel.

3.4   Corn Ethanol Technology

The emergence of corn as the primary feedstock for US ethanol is the product of its abundance and its political prowess. Corn would likely not be any biochemist's first choice as an ethanol feedstock – the sugars needed for ethanol production are bound together in corn as starch, requiring substantial volumes of enzymes to break the sugar-to-sugar bonds. Sugarcane or sugar beets (with their easily extracted, readily fermentable simple sugars) are a more logical choice from a processing standpoint, but most of the United States lacks a suitable climate for sugarcane production, and sugar beets suffer from a low per-acre yield.

So, in the 1970s, when the federal government sought a source for a domestic alternative to unstable imported petroleum, the copious quantities of fermentable sugars cradled in the countless corn cobs across the American Midwest – combined with the potentially powerful constituency of its millions of farmer growers – made corn the logical choice.

Despite the abundance of feedstock, however, early corn ethanol production was at best a breakeven proposition. Throughout the 1970s and early 1980s, ethanol from corn starch required at least as much fossil energy to produce as there was available energy in the finished product.[17] Early biorefineries were inefficient operations requiring large inputs of heat, water and electricity. Corn farming required large inputs of fossil energy as well. As a result, federal support for ethanol in the 1980s and 1990s was increasingly seen as a handout to farmers.

Some skepticism of bio-ethanol remains, but the corn ethanol industry of today bears little resemblance to the pioneer plants of the 1970s and 1980s. Technological advances have transformed both corn production and ethanol processing.

Agricultural biotechnology and improved cropping practices have doubled per acre corn yields since 1970.[11] Corn farmers have also become more efficient in their use of chemical fertilizers. Since 1985, per acre use of nitrogen, phosphate and potash for corn production have all declined (by 6%, 22% and 39% respectively through 2000), while fertilizer production efficiency has increased.[18]

Even greater efficiency gains have been achieved through the use of industrial biotechnology and other new processing technologies at the biorefinery. Two distinct approaches to corn ethanol production are now in use. Both have benefited from substantial technological improvements.

The first ethanol refineries were established at corn wet mills, where corn is processed into its constituent components for animal feed, oils and sweeteners. Ethanol was seen as a way to add value to the starch stream. In today's wet mills, the corn is steeped in water to begin to break down the starch and protein bonds. The mix is ground and screened to remove the germ and fiber, and the starch is separated from the protein. Enzymes are added to the starch to break it down via hydrolysis into glucose. The glucose is then fermented in the presence of yeast to produce ethanol.

Corn wet mills are capital-intensive operations and must typically produce at least 100 million gallons per year of ethanol to justify the necessary capital investments. Most of the corn wet mill ethanol facilities in the US are operated by large agri-business concerns, led by Archer Daniels Midland Company (ADM), the world's largest ethanol producer. Wet mills have become increasingly efficient over time, but only one wet mill ethanol facility has been constructed in the US since 1997.

Corn dry milling is now the dominant technology for ethanol production in the United States. Over 80% of US ethanol facilities are now dry mills.[19] In dry milling, the corn kernel is first ground into coarse flour and then cooked in water. Enzymes are added to break the starch down into glucose in a process called saccharification, creating a mash, which is cooled and fermented. The alcohol is distilled from the mash and the remaining mash is dried to produce a high value animal feed known as distillers dry grains (DDGs). Some of the nutrient-rich liquid is often added back to create distillers dry grains with solubles (DDGS).

Dry mills are far less capital intensive than wet mills. Modern dry mill capital costs average as little as $1.00 per gallon of annual production capacity,[20] versus approximately $2.00 per gallon capacity for the most recent wet mill facilities.[21] This has allowed many smaller investors to enter the ethanol market. Smaller ethanol producers are also eligible for a 10 cents per gallon tax incentive in addition to the 51 cent-a-gallon VEETC. As a result, approximately half of all ethanol biorefineries are now owned by farmer co-operatives. Nearly all new ethanol plants constructed today are dry mill facilities.

One of the largest operating costs for both wet- and dry-mill biorefineries is the cost of fuel for heat and power. The majority of ethanol biorefineries today are powered by natural gas, but natural gas prices in the US have tripled over the past decade. (Short supplies drove natural gas prices to over 5 times their late 1990s levels in autumn of 2005.)[22] As a result, a growing number of biorefineries are converting to coal power, raising concerns over the greenhouse gas emissions from such facilities.

However, with rapid growth in dry mill construction has come rapid improvement in processing efficiency. Molecular sieves have greatly reduced the cost and energy required for ethanol dehydration. Modern dry mills increasingly recycle process heat and water. Improved enzymes and yeasts have increased ethanol yields from corn more than 20 % since the early 1980s.[10] And a growing number of biorefineries are making use of renewable and waste sources of heat and power by burning animal waste or agricultural biomass residues.

Current dry mill ethanol production now appears to yield approximately 1.3 to 1.4 units of usable energy for every unit of fossil energy input – a 40 % reduction in fossil energy consumed versus gasoline, despite bio-ethanol's roughly one-third lower per gallon energy content. As a result, ethanol from modern dry mills results in 15–40 % fewer greenhouse gas emissions per unit usable energy than gasoline on a lifecycle basis.[23–25]

Two emerging technologies have the potential to further substantially improve the economics and environmental profile of dry mill ethanol production. The newest dry mill designs now incorporate advanced fractionation technology, in which the non-starch components of the kernel are finely ground and separated and directed to higher-value uses, much as they are in wet mills. This approach produces a much higher-value DDG for animal feed and a fiber stream, which can be burned to power the facility, or processed in a cellulosic ethanol unit to further increase ethanol yields (see Figure 3.4).

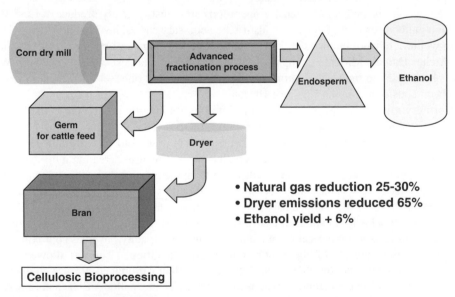

Figure 3.4   *Advanced fractionation processing.*
Source: Biotechnology Industry Organization.

One of the most transformative technologies for dry mill ethanol facilities may be the arrival of 'raw starch' or 'no cook' hydrolysis. Enzyme manufacturers Novozymes and Genencor have both introduced enzymes that can break down corn starch at or near room temperatures rather than the 105–150 degrees Celsius typically required, greatly reducing energy inputs. Some challenges remain in implementing this process, but a growing number of dry mills may soon be adopting this more efficient technology. The combination of new fractionation technologies and 'no cook' enzyme technology can yield approximately a 6 % increase in ethanol yield.

Even with these and other technological advances, the upper limit on sustainable ethanol production from corn remains well below the levels required to establish bio-ethanol as a true alternative fuel in the United States. The greenhouse gas benefits of corn starch ethanol, while better than petroleum, are also likely to remain modest relative to the challenges of global climate change. As a result, a growing coalition of environmentalists, rural advocates and national security groups is instead urging policy makers to invest in ethanol production from cellulosic biomass. The US government is listening.

3.5 Cellulosic Ethanol

January 31, 2006, began as an ordinary day in the United States ethanol industry, but by the time the evening was over the identity of bio-ethanol in the US had changed forever. President George W. Bush in his State of the Union address declared that 'America is addicted to oil' and that to break the addiction, the United States must invest 'in cutting-edge methods of producing ethanol, not just from corn, but from wood chips and stalks, or switch grass'. He set a national goal of practical and competitive cellulosic ethanol within six years. Suddenly, ripples went out through the venture capital community and overnight 'cellulosic' was a household word. The race to the first commercial production of ethanol from cellulose was on.

Scientists have long recognized that cellulosic biomass has tremendous potential to reduce fossil fuel dependence. Agricultural and other residues essentially offer what could be considered a 'free' or much cheaper source of energy. (Since this material is otherwise unutilized, no fossil energy or other inputs are consumed for their production.) Most dedicated energy crops under consideration also require minimal inputs, since the leading candidate species tend to be perennials needing little or no tillage or fertilizer – though some concerns have been raised regarding the consumption of water by fast-growing tree varieties.

Because cellulosic biomass also contains lignin, a fibrous material with similar energy qualities to coal, cellulosic biorefineries are expected to be energy self-sufficient, with the production of cellulosic ethanol powered by the burning of lignin. Thus, little or no fossil energy is needed for processing.

As a result, cellulosic ethanol is estimated to provide over 10 units of useful energy for every unit of fossil energy input – a 7-fold improvement over current corn starch ethanol production and more than 12 times better than gasoline. Net greenhouse gas emissions from cellulosic ethanol are expected to be on average 85 % lower than gasoline. Cellulosic ethanol from switchgrass, a species of perennial grass native to the American prairie that sequesters carbon in the soil as it grows, may even provide a net greenhouse gas benefit to the atmosphere, reducing greenhouse gas emissions by over 100 % versus gasoline. [23–25]

Scientists have tried for decades to overcome the recalcitrance of cellulose and to make cellulosic ethanol commercially using high pressure, steam or strong acids to free cellulose sugars. The federal government began investing in cellulosic ethanol in the 1970s, but nearly three decades of research yielded only modest progress towards commercially viable production. Because plants produce lignocellulose to provide structural integrity, biomass has proven a challenging substrate.

In the late 1990s, scientists turned to a new technology for breaking down the lignocellulosic structure. Industrial biotechnology, as this emerging field has come to be known, makes use of the molecular machines that break down lignocellulose in nature, and improves their effectiveness using an array of genetic tools originally developed to treat diseases or improve crop varieties.

This biochemical approach makes use of cellulose-slicing proteins known as cellulase enzymes, found in the intestines of termites or jungle-dwelling fungi, to isolate simple sugars. The first big breakthrough in this technology came in 2005, when, under US DOE contract administered by the National Renewable Energy Laboratory (NREL), enzyme manufacturers Novozymes and Genencor announced that they had dramatically reduced the cost of biochemical production of cellulosic ethanol. Until the companies' announcements, cellulase enzyme costs were generally acknowledged to be over $5.00 per gallon of ethanol, making commercial biochemical production of cellulosic ethanol in a market of $2.00 per gallon gasoline impossible. By 2005, both companies had developed genetically enhanced microbes capable of producing cellulase enzymes for less than 20 cents a gallon. Commercial production of cellulosic ethanol was within reach (see Figure 3.5).

Several technological challenges to commercial cellulosic ethanol production remain. Current estimates of cellulosic ethanol production costs range from under $2.00 a gallon to

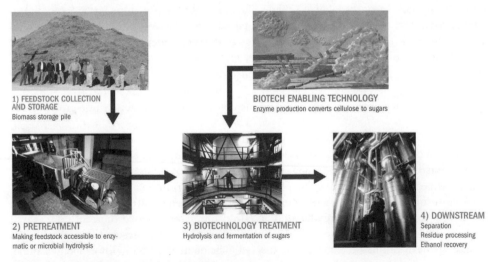

1) FEEDSTOCK COLLECTION AND STORAGE
Biomass storage pile

BIOTECH ENABLING TECHNOLOGY
Enzyme production converts cellulose to sugars

2) PRETREATMENT
Making feedstock accessible to enzymatic or microbial hydrolysis

3) BIOTECHNOLOGY TREATMENT
Hydrolysis and fermentation of sugars

4) DOWNSTREAM
Separation
Residue processing
Ethanol recovery

Figure 3.5   *Biochemical production of cellulosic ethanol.*
Source: Biotechnology Industry Organization.

Remaining Technology and Cost Barriers

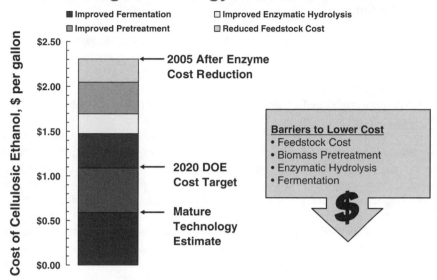

Figure 3.6 *Remaining technology and cost barriers to cellulosic ethanol (p. 23 of 35).*
Source: National Renewable Energy Laboratory.

over $3.00 a gallon, with a US Department of Energy estimate of $2.25 a gallon.[26] Efficient collection, transport and storage systems for cellulosic biomass must still be developed. More efficient pre-treatment technologies are needed to improve biomass receptivity to cellulase enzymes. There is still room for improvement in cellulase enzyme cocktails. And more effective fermentation organisms are needed to convert both the pentose (5-carbon) and hexose (6-carbon) sugars present in cellulosic biomass (see Figure 3.6).

The US Department of Energy estimates that with these technological advances, the cost of cellulosic ethanol production could fall below $1.00 per gallon, making it less expensive than gasoline or today's corn starch ethanol.[26]

In addition to the biorefinery grants issued by the DOE, a growing number of cellulosic biorefinery projects are proceeding without federal assistance. Massachusetts-based Celunol Corporation recently broke ground in Louisiana on a pilot-scale biorefinery to make ethanol from bagasse, the residue from sugarcane processing. Another Massachusetts-based company, Mascoma, is building a $30 million plant in upstate New York to make ethanol from wood chips.

Several companies are also pursuing bio-ethanol production via biomass gasification. In gasification, the biomass is partially combusted and the resulting gases are converted biologically or chemically to ethanol. As petroleum prices have risen, gasification has become a viable route to ethanol production, particularly for low-cost mixed biomass sources such as municipal solid waste. But gasification is a relatively mature technology with substantially less opportunity for cost improvement. Federal investment in gasification has been limited, and is expected to continue to emphasize biochemical processing.

3.6 The Future

While the US ethanol industry is currently enjoying halcyon days, much must still be accomplished if the United States is to meet the President's goal of 60 billion gallons of biofuels production and 30 % displacement of petroleum by 2030.

Effects of ethanol production on food and feed prices may continue to be a concern as starch ethanol production expands. Short-term price increases at various points along both the food and fuel value chains can be expected until markets stabilize.

Producers of corn and ethanol will need to continue to work to ensure that production minimizes environmental impacts, particularly with respect to fertilizer run-off, which is suspected to have contributed significantly to hypoxia and eutrophication in the Gulf of Mexico. A growing number of aquifers in the Midwest are thought to be under stress from water consumption for corn irrigation and ethanol production, highlighting the importance of efficient use and recycling of resources.

Sustainable methods of production and collection must be adopted for all biomass crops as cellulosic ethanol production evolves. Minimum-till or no-till cropping practices are gradually being adopted, and will need to be increased in order to reduce soil runoff if residues are to be collected.[14] Sustainable production systems for switchgrass and other potential dedicated energy crops must also be developed in conjunction with agricultural producers.

Nationwide, substantial investments in ethanol infrastructure are also needed. At the farm level, one-pass harvesting equipment must be developed for collection of crop residues. Advanced soil carbon models must be available to farmers to determine how much residue can be sustainably collected for use as biorefinery feedstock. Farmers transitioning to dedicated energy crops will also need assistance in establishing their crops, since many dedicated energy crops will likely require several years' growth before harvesting can begin.

Expanded regional rail networks will be needed to transport cellulosic feedstocks – or liquid sugars, if processing of biomass occurs at the farm level. And upgrades to existing pipeline networks – or dedicated ethanol pipelines – will be needed to transport ethanol to urban markets.

Biorefineries will also need to evolve to incorporate principles of industrial ecology and to produce more than just ethanol. Just as today's petroleum refineries produce a wide range of chemicals and other products in additional to gasoline, future biorefineries must evolve to extract high-value chemical co-products as well. Integrated biorefineries that produce a range of biofuels, bio-chemicals and biobased products will drive down the cost of bio-ethanol and better ensure its success in the transportation marketplace.

Federal research and assistance in commercializing cellulosic biofuels will continue to be needed. Research to improve enzyme technology will need to be conducted in parallel with commercialization, and pioneer cellulosic ethanol producers will need some assurance that a market for cellulosic ethanol will exist no matter what happens with unstable oil prices. Tax incentives or other incentives to fuel blenders will be needed to ensure that the first commercial cellulosic ethanol makes its way into the fuel supply.

Finally, as the bio-ethanol industry grows, federal and private investment in fundamental biofuels research must continue. New enzyme cocktails and more efficient fermentation organisms are needed for cellulosic feedstocks. And more research is needed to develop higher-order alcohols with greater energy densities, such as the recent alliance between BP and DuPont to commercialize bio-butanol.

There is much in recent US policy discussions to suggest that the federal government will be willing participants in many, if not all, of these areas. Discussions are now underway on a rewrite of federal agriculture policy. The most recent US farm bill is due to expire in 2007, and leaders in both the House of Representatives and the Senate have indicated that biofuels programs – particularly investments in cellulosic feedstocks – will be a major component of the next farm bill.

Recent legislation introduced in Congress also calls for a greatly expanded Renewable Fuels Standard, with a growing proportion of fuels coming from cellulosic feedstocks. Momentum is also building for a low-carbon fuels standard that would require graduated reductions in the greenhouse gas emissions profile of the nation's fuel supply over time. Discussions also continue on a mandatory economy-wide greenhouse gas cap-and-trade system, both of which would strongly favor cellulosic ethanol. (A growing number of states have already implemented regional climate cap-and-trade systems of their own, though few currently apply to transportation fuels.)

All of this suggests that the US bio-ethanol industry is likely to continue to aggressively expand in the coming decade. The possibility remains that, as in the decades following the Second World War and the oil shocks of the 1970s, petroleum prices may fall to a level where bio-ethanol is priced out of the marketplace, sapping public and federal support for biofuels. But sentiment is growing that increasing demand from China and other developing economies is likely to maintain pressure on fuel prices. With new policy tools such as an expanded renewable fuels standard, and growing pressure to regulate greenhouse gas emissions, expansion of the US biofuels market may be here to stay.

References

1. B. Kovarik, Henry Ford, Charles F. Kettering and the Fuel of the Future, *Automotive History Review*, **32**, 7–27, http://www.radford.edu/~wkovarik/papers/fuel.html
2. US Department of Energy, Energy Information Administration, Energy Kid's Page, http://www.eia.doe.gov/kids/ history/timelines/ethanol.html
3. Wikipedia.com, http://en.wikipedia.org/wiki/1973_oil_crisis
4. US Environmental Protection Agency, press release, Nov. 28, 1973, http://www.epa.gov/history/topics/lead/03.htm
5. Renewable Fuels Association, http://www.ethanolrfa.org/industry/statistics/#A
6. Renewable Fuels Association, http://www.ethanolrfa.org/industry/locations/
7. USA Today, Jun 21, 2006, http://www.usatoday.com/money/industries/energy/2006-06-21-ethanol-usat_x.htm
8. Statement of Keith Collins, Chief Economist, US Department of Agriculture, Before the US Senate Committee on Agriculture, Nutrition and Forestry, Washington, D.C., Jan 10, 2007, http://www.usda.gov/oce/newsroom/congressional_testimony/Collins_011007.pdf
9. EthanolMarket.com, http://ethanolmarket.aghost.net/
10. R. Schnefp, Agriculture-Based Renewable Energy Production, *Cong. Res. Serv.*, Jan. 8, 2007, Order Code RL32712

11. National Corn Growers Association, *How Much Ethanol Can Come from Corn?*, Washington, D.C., 2006, http://www.ncga.com/ethanol/pdfs/2006/HowMuchEthanolCan%20ComeFrom Corn.v.2.pdf
12. R. D. Perlack, L. L. Wright, et al., Oak Ridge National Laboratory, *Biomass as Feedstock for a Bioenergy and Bioproducts Industry: The Technical Feasibility of a Billion-Ton Annual Supply*, ORNL/TM-2005/66, Oak Ridge, Tenn., April 2005, http://feedstockreview.ornl. gov/pdf/billion_ton_vision.pdf
13. N. Greene, Natural Resources Defense Council, *Growing Energy: How Biofuels Can Help End America's Oil Dependence*, Washington D.C., December 2004, http://www.nrdc.org/air/ energy/biofuels/biofuels.pdf
14. J. Hettenhaus, Biotechnology Industry Organization, *Achieving Sustainable Production of Agricultural Biomass for Biorefinery Feedstock*, Washington D.C., November 2006.
15. National Ethanol Vehicle Coalition, *How Many E85 Compatible Vehicles Are on American Roads Today?*, http://www.e85fuel.com/e85101/faqs/number_ffvs.php
16. National Ethanol Vehicle Coalition, *E85 Refueling Locations By State*, http://www.e85refueling .com/states.php
17. Ethanol Study Committee, *Ethanol Production From Biomass With Emphasis on Corn*, University of Wisconsin, Madison, September 1979.
18. H. Shapouri, J.A. Duffield and M. Wang, *The Energy Balance of Corn Ethanol: An Update*, USDA, July 2002.
19. Renewable Fuels Association, *Ethanol Industry Outlook 2007: Building New Horizons*, Feb. 2007, http://www.ethanolrfa.org/objects/pdf/outlook/RFA_Outlook_2007.pdf.
20. R. Dale and W. Tyner, *Economic Analysis of Dry-Milling Milling Technologies*, Power-Point Presentation, Purdue University DOE Grant # DE-FG36-04G014220, http://cobweb.ecn. purdue.edu/~lorre/16/Midwest %20Consortium/9-tyner-dry-mill-ethanol-model-nov17.pdf
21. S. Eckhoff, *Wet Milling Versus Dry Grind: Effect of Co-Products and Economy of Scale*, Presentation at Sustainable Bioenergy: Focus on the Future of Biofuels and Chemicals, April 13-14, 2006, UIUC, http://www.sustainablebioenergy.uiuc.edu/Presentations/Eckoff.pdf.
22. US Department of Energy, Energy Information Administration, *U.S. Natural Gas Wellhead Price History*, http://tonto.eia.doe.gov/dnav/ng/hist/n9190us3m.htm.
23. M. Wang, Argonne National Laboratory, *Energy and Greenhouse Gas Emissions Impacts of Fuel Ethanol*, presentation to the National Corn Growers Association Renewable Fuels Forum, Washington, D.C., August 23, 2005, http://www.anl.gov/Media_Center/News/ 2005/NCGA_Ethanol_Meeting_050823.ppt.
24. P.R. Adler, S.J. Del Grosso and W.J. Parton, Life-Cycle Assessment of Net Greenhouse-Gas Flux for Bioenergy Cropping Systems, *Ecological Applications*, **17(3)**, 675–691 (2007).
25. A.E. Farrell, R.J. Plevin, B.T. Turner, A.D. Jones, M. O'Hare and D.M. Kammen, Ethanol Can Contribute to Energy and Environmental Goals, *Science,* **311**, 506–508 (2007).
26. Statement of Michael Pacheco, Director, National Bioenergy Center, National Renewable Energy Laboratory, to the House Committee on Energy and Commerce, Washington, D.C., May 24, 2006, http://energycommerce.house.gov/reparchives/108/Hearings/05242006hearing1909/ Pacheco.pdf

4

Bio-Ethanol Development(s) in Brazil

Arnaldo Walter

Department of Energy and NIPE, State University of Campinas (Unicamp), Brazil

4.1 Overview

Brazil has consumed ethanol as automotive fuel, in large-scale, for more than 30 years. The country is currently the second largest world producer of ethanol (was surpassed by USA in 2006) and is the only country where biofuels are strictly competitive with oil derivatives. Brazil currently produces more than 18 billion liters (Gl) of ethanol from sugarcane and exports more than 3.5 Gl; production shall reach 28–30 Gl and at least 6 Gl shall be exported by 2013. A deep reduction of production costs was achieved due to technology development (both in the agriculture and in the industry) and up scaling of production units. Ethanol production in Brazil is currently much more sustainable than in the past, but there are still room for improvements: the best results are related with reduction on GHG emissions and developments are necessary regarding biodiversity preservation and social aspects. Anyhow, many Brazilian producers are currently in condition to sell to markets where sustainability of biofuels production is an important issue. In order to keep on competitive, a big challenge for Brazil is regarding continuous technology development, with focus on diversification of the production process (e.g. development of hydrolysis of sugarcane bagasse) and of the products (e.g. enlarging electricity production and producing chemicals and other materials). Brazilian experience on large-scale production of ethanol is really a good case study and the country can contribute a lot in order to make ethanol a real commodity.

Biofuels Edited by Wim Soetaert and Erick J. Vandamme
© 2009 John Wiley & Sons, Ltd

4.2   Introduction

The Brazilian experience in the large-scale production of biofuels is essentially regarding ethanol, as the biodiesel program was created in 2004 and no important result has been achieved so far. On the other hand, large-scale ethanol production as automotive fuel started in 1975 and since then the ethanol production has increased more than 28 times. In 2006 ethanol production (anhydrous + hydrated) has reached 17.8 billion liters (Gl).[1] During almost 30 years Brazil was the largest producer of fuel ethanol in the world but is currently the second largest (the production of USA in 2006 reached almost 18.4 Gl).[2]

All ethanol production in Brazil is based on sugarcane. There are two main types of distilleries used in the production of sugarcane-based ethanol: annexed and autonomous. An annexed distillery is built alongside a sugarcane mill and can provide considerable flexibility against price fluctuations; is currently the preferred option in Brazil: a typical mill can vary the share of sugarcane crushed for ethanol production from 40 % to 60 % regarding the total. In an annexed distillery the feedstock is usually a blend of cane juice and molasses. On the other hand, in an autonomous distillery the prime aim is the production of ethanol. This is only justified where there is a large and highly secure market for ethanol, as in the case of Brazil, but through the years most of the existing autonomous distilleries have been converted to annex (it is estimated that there are less than 50 autonomous distilleries in almost 350 units). In Brazil an autonomous distillery just uses cane juice.[3] The synergy of the production of sugar and ethanol in the same mill is one of reasons for the low costs of ethanol production in Brazil. This has been called the 'Brazilian model' on ethanol production.

In Brazil ethanol is either used as oxygenate and octane enhancer of gasoline or in neat-ethanol engines. In the first case, anhydrous ethanol (99.6 Gay-Lussac) is blended to gasoline in the proportion of 20–26 % in volume of ethanol. All gasoline sold in Brazil is in fact gasohol. Hydrated ethanol (95.5 Gay-Lussac) is used in neat ethanol vehicles since late 1970s and, since 2003, in flex-fuel vehicles – FFVs.

Currently ethanol covers more than 30 % of the energy demand of the automotive transport (energy basis) and the forecasts are that ethanol use will remarkably rise in the years to come. In Brazil ethanol is competitive vis-à-vis gasoline and no subsidies have been applied since late 1990s.

4.3   The Brazilian Experience with Ethanol

In Brazil, ethanol has been blended with gasoline since the 1930s, but it was just in 1975 that the Brazilian Alcohol Program (PROALCOOL) was created aiming at partially displacing gasoline in automotive transport. At that time the country was strongly dependent on imported oil (90 % in 1973 and more than 80 % in 1975) and gasoline was the main oil derivative consumed (more than 25 % of the demand of oil derivatives – energy basis). During the first years the production of anhydrous ethanol was the priority, in order to be blended with gasoline.

In 1979, during the second oil chock, the Brazilian Government decided to enlarge the Program, supporting large-scale production of hydrated ethanol to be used as neat fuel in modified engines. In the 1975–79 period, ethanol production was accomplished by new distilleries annexed to the existing sugar mills, while in the 1979–85 period many autonomous distilleries were built.

It is estimated that from 1975 to about 1985 about US$ 11–12 billion were invested to create a structure able to produce 16 billion liters of ethanol per year. On the other hand, savings regarding avoided oil imports were estimated as US$ 52.1 billion (January 2003 US$) from 1975 to 2002.[4]

Brazilian government reduced the support to the producers after 1985 and, as consequence, the ethanol market faced difficulties during the 1990s. The crucial moment was an ethanol supply shortage in 1989–90 that induced a strong drop in sales of neat ethanol cars. For instance, sales of neat ethanol vehicles that have reached 92–96 % during the Eighties were continuously reduced until summing up just about 1000 new vehicles per year in 1997–98.[5]

The PROALCOOL finished during the 1990s as long as government support ceased. From the institutional point of view changes started in the early 1990s, first with liberalization of fuel prices to consumers and secondly, in the late 1990s, with full deregulation of sugarcane industry. The positive results started to be noticed in 2001, when due to a larger price difference between ethanol and gasoline, sales of neat ethanol cars increased.

More recently, since the launch of FFVs (March 2003) there is a boom on sales of vehicles able to run powered by ethanol. In mid-2007, the sales of FFVs have surpassed 85 % of the total sales of new automobiles (cars and light vehicles). Currently all car manufacturers based in Brazil have at least one FFV model available and some manufacturers have decided that some models are exclusively FFV. Regarding neat ethanol vehicles, the main advantage of the FFVs is that these engines can operate with any fuel mix between regular gasoline (in Brazil, E20-E25, i.e. 20–25 % of anhydrous ethanol in the fuel blend) and pure hydrated ethanol (E100).

The ethanol market in Brazil has started a new momentum due to the success of FFVs. Close to the ethanol production regions, and during most of the year, the price of hydrated ethanol at service stations is much lower than the price of gasoline C (the blend of pure gasoline and anhydrous ethanol). Figure 4.1 shows the price ratio between gasoline C and hydrated ethanol in the city of São Paulo, from February 1999 (beginning of full deregulation at the fuel's market) to December 2006. For FFVs the breakeven ratio varies depending on the car model and on the traffic conditions but, in general, it is economic to run with hydrated ethanol if its price is lower than 70 % of the price of gasoline C (volume basis).

Figure 4.2 shows the evolution of ethanol production in Brazil from 1970 to 2006. Figure 4.3 shows the share of the energy consumption (cars and light vehicles) covered by different fuels in the period 1970–2005. In Brazil cars and light vehicles cannot use diesel oil. Natural gas consumption by automobiles has been deployed since late 1990s, especially in cabs and companies' vehicles, as a way to accelerate the return on investments in the Bolivia–Brazil pipeline. Due to constraints on natural gas supply, this market will not grow in the years to come. In mid 1980s ethanol covered more than 50 % of the energy consumption of automobiles in Brazil and currently covers slightly more than 30 % of the fuel consumption.

The technology of FFVs and the relative low price of ethanol vis-à-vis gasoline are the reasons why it is predicted that the domestic market shall reach 28–30 Gl by 2013 and possibly 45–50 Gl by 2030.[9,10] As of 2006 Brazil had 335 industrial units producing

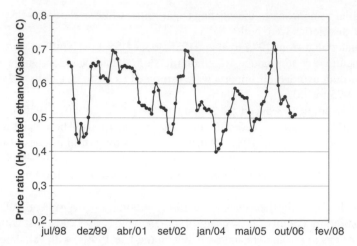

Figure 4.1   *Price ratio between hydrated ethanol and gasoline C in the city of São Paulo.*
Source: [6,7].

ethanol and about 80–90 under construction. By the end of 2006 the installed capacity of production was 18 Gl of ethanol. Total sugarcane production in 2006 was 425 million tonnes (roughly 50 % used for sugar production and 50 % used for ethanol production) and shall reach 730 million tonnes by 2013. According to the forecasts, sugarcane production shall surpass 1000 million tonnes between 2020 and 2025.

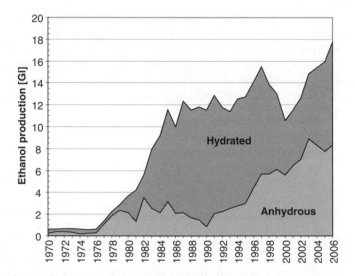

Figure 4.2   *Evolution of ethanol production in Brazil from 1970 to 2006.*
Source: [8].
Notes: 1975 – beginning of the Ethanol Program; 1979 – beginning of hydrated ethanol use as fuel; 1985 – reduction of the support of Federal Government to producers; 1990 – shortage on ethanol supply; 1999 – deregulation of sugarcane industry; 2003 – first FFV in the market.

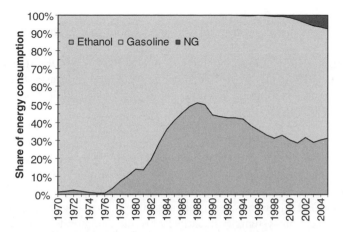

Figure 4.3   *Share of energy consumption by automobiles in Brazil.*
Source: [8].

Since 2000 (and mainly since 2004) exports of fuel ethanol have increased. Total exports from Brazil in 2003 were about 750 Ml but reached 3.4 Gl in 2006 (1.75 Gl straight to USA, more about 480 Ml to USA through Caribbean and Central America countries, 580 Ml to the European Union and 270 Ml to Japan).[1] A conservative estimate is that exports shall reach 7 Gl in 2013.

Brazil has about 30 % of the sugarcane planted area in the world.[11] As of 2006 the planted area with sugarcane was 6.4 Mha, about 2.5 % of the agricultural land available (263.6 Mha in 2003, according to[11]) and approximately 10 % of planted areas (66.6 Mha in 2003[11]). There is a huge concentration of sugarcane production in the Center-South region of Brazil (87 % of the total production), and mainly in the state of São Paulo (62 % in 2006, both for sugar and ethanol production).[1] The current growth of sugarcane production is occurring in São Paulo and in other states in the Center region; just a few new mills have been built in the Northeastern region. Expansion has been almost exclusively based on replacement of other crops (e.g. orange) and pastures.

4.4   Policy and Regulatory Instruments Applied to Deploy Large-Scale Ethanol Production

The chicken and egg problem is classical for alternative fuel vehicles: who will buy them if a fuelling infrastructure is not in place, and who will build the infrastructure if there is no vehicles in the market?[12] It is necessary to reduce risk perception both for producers and consumers, and this is one of the main challenges for deploying an energy source.

During the first 15 years of Brazilian ethanol program, supplies and demands were both stimulated and adjusted through central coordination. Producers accept the Program since the very beginning, because the Program in itself was also created in order to minimize the frequent difficulties faced by sugarcane sector due to the excess of sugar production and fluctuations of its international prices. In addition, the required investment was assured by credits given at low interest rates and risks were extremely reduced as off-taking guarantees were given at fixed prices – both to sugarcane and to ethanol. In fact, fixed prices for producers and consumers played an essential role in the general trust of the program.[13]

Also aiming at assuring the supply the government has obliged the state-controlled oil company (PETROBRAS) to provide and operate the required infrastructure of transport, storage, blending and distribution. Eventual losses of ethanol commercialization were also assumed by PETROBRAS.

In parallel, in order to induce the consumption, the government negotiated with the automobile industry (at that time, four main companies were based in the country) to introduce the required modifications in engines and parts. More modifications are required as large is the share of ethanol in the fuel blend; for instance, for 25–100 % ethanol in the fuel blend, modifications include materials substitution (e.g. of the fuel tank, fuel pump, electronic fuel injection system) and new calibration of devices (e.g. of ignition and electronic fuel injection systems).[5] In early 1980s, the automobile industry has accepted to give full warranties to the consumers. The R&D efforts regarding engines able to run with blends and straight ethanol started at a federal research center (Aeronautics Research Center) where engines development and tests were performed. The first neat ethanol engine was commercially available in 1979 and technology was quickly transferred to the automobile industry.[14]

On the other hand, the ethanol market was induced by mandates. In 1975 a mandate for 20 % anhydrous ethanol (E20 – volume basis) on fuel blend was established. However, in reality just by early 1980s the share of ethanol into all gasoline commercialized reached 20 %. Through the years the share of ethanol in fuel blend has changed, as can be seen in Figure 4.4. The ethanol share was reduced to 13 % between 1989 and 1993, during the supply ethanol crisis, while in 1993 it was defined by law that the share of ethanol in fuel blend should be in the 15–25 % range, depending on the conditions of ethanol supply. Since them the lowest level reached was 20 %. In reality, this relatively wide range allows

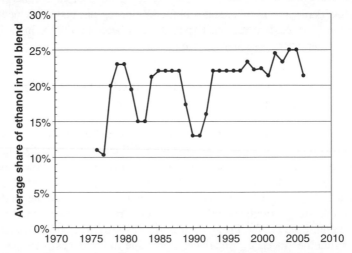

Figure 4.4   *Average share of ethanol (anhydrous) in the fuel blend, according to mandates.*
Source: [15].

the production be shifted to more sugar if necessary, allowing the producer to maximize its earnings. Currently, the share of anhydrous ethanol in the fuel blend is 25 %.

Moreover, consumers have been stimulated to buy neat-ethanol cars through lower taxes regarding those applied over gasoline vehicles. In addition, due to the control over fuel prices that existed until the mid 1990s, it was imposed that ethanol prices to consumers would be close to 65 % of the price of gasoline (volume basis). The fleet of neat-ethanol vehicles had reached 4.5–5 million in the mid 1990s.

In Brazil, taxes have a strong impact over the fuel price to consumers. Currently, six different taxes and contributions have been applied over automotive fuels, being just one equivalent to the value-added tax (VAT). In 2005 the average taxation over gasoline C in Brazil was estimated as 47 %, while the average taxation over hydrated ethanol was evaluated as 34 %. In addition, in São Paulo state (the largest producer and consumer of ethanol in Brazil), the taxation on hydrated ethanol was close to 20 % in the same year.[16] Also, in the state of São Paulo an additional advantage is the lower value of the annual license paid by owners of neat-ethanol vehicles (including FFVs).

Direct subsidies were completely eliminated with the deregulation process that finished in the early 2000s. However, a tax exemption policy is in place and part of the benefits received by ethanol consumers is due to lower taxes applied to ethanol regarding those paid by gasoline consumers. Anyhow, it should be noticed that in Brazil the taxation applied to diesel oil is even lower than the correspondent applied to ethanol (about 27 % in 2005, on average).[16]

4.5   Cost Reductions

Brazil has the lowest production cost of ethanol in the world and is so far the only country where biofuels are strictly competitive vis-à-vis oil derivatives. Figures about production costs of ethanol in Brazil vary due to the set of mills considered and also according to the exchange ratio used; the exchange ratio of Brazilian currency (Real) vis-à-vis the main foreign currencies has declined during the last years and thus, so have the ethanol costs expressed in Euro or American dollars, for instance.

Ethanol production costs fell on average 3.2 % per year in the South–South region since 1975 and about 1.9 % per year in the Northeastern region.[17] In 2001 it was estimated that the production cost of hydrated ethanol in a mill with good performance was around R$ 0.45,[18] or about US$ 0.18 per liter, considering the exchange rate at that time. In a comparative study published in 2004[19] it was evaluated that the average production cost of anhydrous ethanol in Brazil at that time was 0.145 Euro per liter, or US$ 0.18–0.19/litre. Production cost of anhydrous ethanol is about 5–10 % higher than the cost of hydrated ethanol.

Table 4.1 compares production costs of anhydrous ethanol in Brazil to the costs in USA and Germany. It can be seen that comparing production costs per se, i.e. without subsidies, the average cost in USA is more than twice higher than in Brazil, while the production cost in Germany is more than three times higher. The same conclusion was presented by Worldwatch Institute[20] regarding relative costs, based on information from the International Energy Agency and the US Department of Energy: 14–20 Euro/m^3 of

Table 4.1 Production costs of ethanol in Brazil, USA and Germany (Euro/m^3)

	Brazil	USA	Germany	Germany
Feedstock	Sugarcane	Corn	Wheat	Sugar beets
Building	0.21	0.39	0.82	0.82
Equipment	1.15	3.40	5.30	5.30
Labor	0.52	2.83	1.40	1.40
Insurance, taxes and others	0.48	0.61	1.02	1.02
Feedstock	9.80	20.93	27.75	35.10
Other operation costs	2.32	11.31	18.68	15.93
Total production cost	14.48	39.47	54.97	59.57
Sale of by-products		6.71	6.80	7.20
Government subsidies		7.93		
Net production cost	14.48	24.83	48.17	52.37

Source: [19].

anhydrous ethanol in Brazil, 23–35 Euro/m^3 in USA (production from corn) and 28–46 Euro/m^3 in Europe (from grains).

Based on prices paid to producers from 1980 to 2002 (which was a good indicator of the trend of production costs during that period) Goldemberg et al.[4] estimated the ethanol experience curve for the Brazilian production. The progress ratio (i.e. the ratio of unit costs after a doubling of cumulative production to the unit costs before the doubling) in the 1980–85 period was 0.93, and fell to 0.71 in the 1985–2002 period. Addressing the same subject in a more detailed analysis, van den Wall Bake[13] showed that the experience curve of ethanol production in Brazil is a better estimate with a progress ratio of 0.79 over the period 1975–2004, when approximately six cumulative doublings of ethanol production were observed. Industrial processing costs declined more than agricultural costs, but with a lower impact to the overall cost reduction as feedstock represents 60–65 % of the total production costs.

The largest share of the total feedstock cost reduction was due to the development of new varieties of sugarcane (see next section) with indirect impacts on costs of soil preparation, planting, stock maintenance and land rents (due to the higher number of the cuts – five to six – and to larger yields).[13] On the other hand, industrial processing costs were reduced more due to economies of scale, with impacts on investments and on operation and maintenance costs. Furthermore, up scaling lead to vertical chain integration that indirectly allowed optimization of the production chain.[13]

Based on the calculated progress ratios, and assuming that an annual production growth of 5–8 % (has been almost 9 % in over the period 2000–06), total ethanol production cost in Brazil is expected to be reduced about 20 % up to 2015.[13] The prediction presented by[20] is that after 2010 ethanol production costs in Brazil can be further reduced by 10–15 % regarding those verified over 2004–05.

4.6   Technological Development

Comparative advantages of ethanol production in Brazil are mostly due to the technological developments that have been conducted for many years in private companies, research centers and universities. More oriented R&D efforts started in the 1970s with adaptation and optimization of technologies from other sugar and ethanol producing countries, and further with the development of technologies more suited to the local conditions.

Sugarcane research started in Brazil in the early 1930s as a consequence of destruction of plantations due to virus attack. Later, in early 1980s, a more ambitious program aiming at developing new varieties was developed due to the catastrophic effects of diseases (consequence of the enlarged planted area and availability of very few varieties). The R&D efforts were later focused on the enlargement of sugar and ethanol productivity.

In fact, diversification of cane varieties is part of the pest and disease control strategy. Currently there are more than 500 commercial varieties of sugar cane. The top 20 occupy 80 % of the total cane area and the leading variety occupies only 13 %. The duration of use for each variety is becoming increasingly shorter, and at the same time, the number of varieties in use at any given time has been growing. [21]

Agricultural yields and the amount of sucrose in the plant have a strong impact on costs of sugarcane products. Agricultural yields depend on soil quality, weather conditions, agricultural practices and are also strongly influenced by agricultural management (e.g. planting and harvesting timing and choice of sugarcane varieties). The average productivity in the largest area of sugarcane production (Centre-South Region) is around 84 t/ha in a five-cut cycle, but it could be as high as 110–120 t/ha in the state of São Paulo; [22] 87 t/ha is the current average figure in state of São Paulo. [23] Sugarcane cultivation in Brazil is based on a ratoon-system, i.e. after the first cut the same plant is cut several times on a yearly bases. [24] It is worth mentioning that these yield figures cannot be strictly compared with yields reached in other countries as sugarcane cultivation in Brazil is done without irrigation. Since 1975 yields have grown almost 60 % due to the development of new varieties and to the improvement of agricultural practices.

The development of cane varieties also aims to increase the sugar content in the sugarcane – which is expressed by the total reducing sugars (TRS) index. The TRS impacts both sugar and ethanol production and, for this reason, is considered on the sugarcane payment. To give an idea of the evolution achieved, in 25 years TRS almost double and best practice figures are close to 15 %. [5] The combination of higher yields and higher TRS result in lower land use: e.g. to obtain the production of 425 million tonnes (2006) with the productivity of 1975 (about 50 t/ha), more 3 Mha would be necessary.

The largest R&D program in the world regarding genetic development of sugarcane varieties was conducted in Brazil by CTC (former Copersucar Technology Center, currently Sugarcane Technology Center). The main targets of the program were the increase of the TRS, the development of disease-resistant varieties, better adaptation to different soils and the extension of the crushing season. [25] Still regarding improvements on sugarcane varieties, it is worth mentioning the Sugarcane EST Project – SUCEST, which started in 1999. As a result, by the end of 2003 more than 90 % of sugarcane genes were identified.

In addition, many of the results achieved in sugarcane agriculture are due to the introduction of machinery for soil preparation and soil conservation. Machinery was introduced in many operations during the last 30 years, but advances on harvesting are more recent. Currently, in the Centre-South region mechanized harvest is applied over about 40 % of the sugarcane planted area, being 25–30 % harvested without previous burning of the field. [23] Sugarcane is usually burned in the field to allow higher throughput during manual harvesting (without previous burning, the costs of manual harvesting would be about three times higher); during the burning process leaves and tops of the plant are almost completely eliminated, with almost no impact in the sugarcane plant. Due to the lower costs, in some regions up to 90 % of sugarcane is mechanically harvested. It is estimated that with mechanized harvesting costs can be reduced 30 %, but investments required are still very high.

In the years to come, and mainly due to environmental constraints, a gradual growth on green cane harvesting is forecast, i.e. mechanized harvesting without previous field burning. Green cane harvesting will allow the recovery of sugarcane trash (leaves and tops of the plant) and a significant increase in biomass availability (about 30 % in mass basis and about 50 % in energy basis, considering that 50 % of the trash can be transported to the mill). In the state of São Paulo legislation obliges a gradual growth on mechanized harvesting and full mechanization will be reached just in 2017. On the other hand, in most of the Northeast region mechanized harvesting cannot be extensively applied due to topographic conditions.

Green cane harvesting requires the development of more adequate machines, considering local topography and the way sugarcane is planted. The main aspects to be observed are high performance in sugarcane recovery, lower costs, reduction of the amount of soil transported and low level of sugar losses.[25] Appropriate harvesting machines have been developed in Brazil.

Gains on productivity and cost reductions were also achieved due to the introduction of operation research techniques in agricultural management and to the use of satellite images for varieties identification in planting areas. Similar tools have been used in decision-making regarding harvesting, planting and application rates for herbicides and fertilizers.

On the other hand, regarding the industrial process of ethanol production different priorities were defined along the years.[26] In a first moment the focus was put on increasing equipment productivity, eventually with reducing conversion efficiencies. The size of Brazilian mills also increased and nowadays milling capacity is 2–5 times higher than 25–30 years ago;[23] the current standard mill has crushing capacity of 2–3 million tonnes of sugarcane per year. In a second moment the focus was moved to improvements on conversion efficiencies, effort that still continues. Since mid-1980s the conversion efficiency at the industry has grown from 73 to 85/t of sugarcane processed, or 1.6 to 1.9 GJ/t (based on the LHV of anhydrous ethanol – 22.3 MJ/liter). It is expected that conversion efficiency can reach 91/t in 8 years and 92.5/t in 18 years. Finally, during the last 15 years the focus has been on better management of the processing units.

A summary of the main technological improvements in the industrial process is presented in Table 4.2.

Due to the technological developments achieved both on the agriculture and on industry sides, average production yields have grown from 3000 liters/ha.year (67 GJ/ha/yr) in early 1980s to 6500 liters/ha.year (145 GJ/ha/yr) in 2005.[9] Production yields based on conventional process can reach 8000 liters/ha.year (178 GJ/ha/yr) in about 8 years or even 9000 liters/ha.year (about 200 GJ/ha/yr) in case ethanol production from hydrolysis of sugarcane bagasse reaches a commercial stage.

Sugarcane bagasse is a prime candidate for ethanol production via hydrolysis because of its low opportunity cost (below 1 Euro/GJ) and the fact that the existing mill infrastructure can be used. Brazilian companies have an R&D program aiming at developing ethanol production through acid hydrolysis (ethanol and diluted acid as solvent).[3]

Table 4.2   *Main technological improvements in the industrial process*

Process step	Actions	Average and best practice results
Juice extraction	Rise on crushing capacity; Reduction of energy requirements; Rise on the yield of juice extraction.	Extraction yield has improved from 92 up to 97.5 %. Average yield around 96 %.
Fermentation	Microbiological control; Yeast selection based on genetics and better yeast selection; Large-scale continuous fermentation, better engineering and better control of process	Fermentation yield has improved from 83 to 91.2 % (best practice 93 %); Production time has decreased from 14.5 to 8.5 hours (best practice 5.0 h); Wine content has improved from 7.5 % to 9.0 % (best practice 11.0 %); Final yeast concentration has improved from 6 to 13 % (% volume); Reduction of about 8 % on ethanol costs due to continuous fermentation and microbiological control.
Ethanol distillation	Improvements on process control	Average yield has risen from 96 % in early 1990s to up to 99.5 % (result also influenced by higher ethanol wine content).
Cane washing	General improvements	Reduction on water consumption; Reduction on sugar losses (2 % down to just 0.2 % in some cases).
Industry in general	Instrumentation and automation	Impact on juice extraction, evaporation, fermentation, crystallization and steam generation.

Sources: [23,26,27].

The final comment in this section is regarding electricity production from sugarcane bagasse, at the mill site. As shown in Figure 4.5, electricity production has increased since late 1980s and, on average, Brazilian mills are producing surplus electricity (the amount that exceeds 12 kWh per tonne of sugarcane crushed, that is the estimated electricity self-consumption of a typical mill) since 1996. Considering the sugarcane production in 2006 (425 million tonnes), about 3.5 TWh of surplus electricity could be produced and commercialized. However, for the same amount of sugarcane produced, but using both sugarcane bagasse and trash as fuels, the potential is 5–7 higher than what has been produced. The constraint for deploying the potential is not technological: the fact that Brazil still has a large hydroelectric potential available is, in practice, one of the constraints. On the other hand, diversification of the production is one of the main strategies to enhance the competition of ethanol production from sugarcane and it is important to deploy the potential of electricity production from sugarcane residues.

4.7   Is the Ethanol Production in Brazil Sustainable?

Among the driving forces for biofuels, one of the most important is the growing concern about sustainability. Expectations with regard to biomass, in general, and biofuels, in

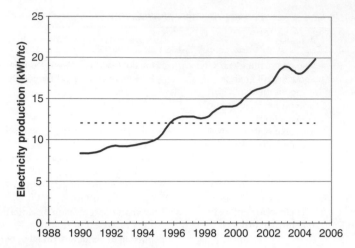

Figure 4.5   *Average electricity production (kWh/t of sugarcane crushed) in Brazilian mills – 1990–2005.*
Source: [1,8].

particular, are high, but there are also risks connected to the production of biomass at large scale.[28] The tendency is that markets, especially in the European Union, claim for certificates and labels in order to – at least – assure the adoption of certain procedures and fulfillment of basic principles along the whole production chain. In a general sense, the main concerns regarding large-scale biofuels productions are associated with the following issues[28]:

1. GHG emissions: biofuels production and use should allow a minimum reduction of GHG emissions, when compared with the use of fossil fuels.
2. Competition with food production: large-scale biofuels production should not jeopardize food production neither contribute to significant raise of food prices.
3. Impacts on biodiversity: large-scale production of biofuels should not impact natural ecological systems nor contribute to the reduction of water availability.
4. Use of pesticides and fertilizers: biofuels production should not cause any important impact on soil and water bodies as a consequence of the large-scale use of agrochemicals.
5. Positive impacts on local economy: biofuels production should positively impact the region where it takes place.
6. Positive impacts on living conditions: biofuels production should positively impact social conditions of the employees and the local population.

In the following text, ethanol production in Brazil is examined according to the six principles described above.

4.7.1   GHG Emissions

One of the strategies outlined by the International Panel on Climate Change (IPCC) is the reduction of greenhouse gases (GHG) emissions by large-scale use of low carbon fuels from

Table 4.3   *GHG emissions during the ethanol life cycle (kg CO_2 equivalent/m^3 of anhydrous ethanol)*

	Average values	Best practices
Productivity (/t)	84.8	92.3
GHG emissions		
Fossil fuels	0.22	0.19
Methane and N2O from trash burning	0.11	0.10
Soil N2O	0.07	0.07
Total GHG emissions	0.40	0.36
Avoided GHG emissions		
Surplus bagasse use[1]	0.14	0.25
Ethanol use[2]	2.86	2.81
Total avoided emissions	3.00	3.06
Net avoided emissions	2.60	2.70

Notes: [1] due to the substitution of surplus bagasse for fuel oil.
   [2] due to the substitution of ethanol for gasoline.
Source: [29].

non-fossil sources, such as ethanol. Reduction of GHG emissions from ethanol depend on many factors, e.g. the raw material and the production conditions. For the production of ethanol from sugarcane under Brazilian conditions, a study by Macedo et al.[29] shows that the input/output energy ratio (i.e. the amount of energy generated as ethanol regarding the input of fossil fuels) varies from 8.3 to 10.2. The same study shows that the avoided emissions of GHG for anhydrous ethanol use are 2.6–2.7 kg CO_2 equivalent/liter of ethanol. Table 4.3 summarizes the main results of this study.

Regarding GHG emissions, the comparison of total emissions of CO_2 equivalent in a 'well-to-wheel' basis (i.e. considering the life cycle of each fuel) is most accepted. In some cases emissions from biofuels are as high as that from gasoline, whereas other combinations of feedstock and conversion processes can reduce 'well-to-wheel' CO_2 emissions to near zero (IEA, 2004). There is a general consensus that using bioethanol produced from sugarcane, according to the Brazilian conditions of production, 80–90 % of the GHG emissions can be reduced in respect to the use of gasoline (emissions per km traveled). As a matter of comparison it should be noticed that the use of ethanol derived from grains (e.g. wheat) brings about 30 % reduction of CO_2 emissions.[30,31] Ethanol from sugarcane (considering Brazilian conditions of production) is the best biofuel alternative concerned with reduction of GHG emissions.

4.7.2   Impacts on Food Production

As previously mentioned, in Brazil sugarcane production occupies about 10 % of the current planted area and about 2.5 % of the agricultural land available. More specifically, ethanol production is responsible for just half of this land use, as just 50 % of the sugarcane is used for ethanol production.

It is estimated that in Brazil the area available for the enlargement of the agriculture, without deforestation, is 110 million hectares. Thus, considering that ethanol production is going to double during the next 5–18 years, the required area would be about 3 % of the land available as far as current overall productivity is considered.

In Brazil it is well accepted that large-scale sugarcane plantation in Brazil has not affected food production and even a substantial enlargement of the ethanol production would be possible without meaningful constraints. For the past 35 years the harvested areas of corn and soybean crops have increased dramatically, while the area harvested for other cultures has remained almost identical. In 2000, the land dedicated for soybeans production was about three times larger than the area occupied with sugarcane, while the land occupied with corn plantations was almost twice as large. [32]

4.7.3   Impacts on Biodiversity

One important aspect to be considered is that due to the land availability in Brazil, potential impacts on biodiversity regarding large-scale production of ethanol can be minimum. Sugarcane is not directly responsible for deforestation in Brazil and the expansion of sugarcane plantations has occurred, displacing other crops and/or using degradable lands previously used for cattle. Environmental legislation clearly specifies that it is forbidden to engage in any type of deforestation. [32] On the other hand, it is very difficult to evaluate indirect effects of sugarcane expansion, as pasture, for instance, can be moving to deforested areas.

Anyhow, there is just one small sugarcane mill in the Amazon region and just another mill was recently considered to be built close to the Amazon region, but not in the area covered by forest. As a matter of fact, both the weather and the quality of soil in the Amazon region are inadequate for sugarcane production. The expansion of sugarcane must mainly occur in the *cerrado* area (the ecosystem in the Center region of Brazil) but, considering the extension of required area, this option is also controversial.

Even with plenty of land available, regulation of land use and the definition of areas suitable for different economic activities are absolutely necessary in Brazil. Federal Government has recognized the importance of such actions and some concrete movements in this regard are expected in the short run.

Given the nature of the farming sector (with high concentration of tenure), in Brazil large track of lands are planted with sugarcane. The legislation obliges that 20 % of the land have to be left aside in order to preserve native vegetation and biodiversity but, unfortunately, law enforcement has not been so far as strict as expected. In a country like Brazil, effective results in this regard require pressure from the most conscious sectors of the society.

Water consumption is another very important topic as far as large-scale production of biofuels is concerned. In Brazil sugarcane is basically planted without irrigation, and this is an important advantage both from an economic and from an environmental point of view. However, water consumption in the industrial process can be significant, and special attention is required in order to reduce water use. In a mill with no recycling, water consumption is higher than 20 m^3/t of sugarcane crushed. Average figures in state of São Paulo are 2.5 m^3/t of sugarcane crushed and the short to mid-term target is to achieve consumption equivalent to 1 m^3/t of sugarcane crushed; some mills still have water consumption as low as 0.7 m^3/t. Considering the milling capacity of some industrial units and the low availability of water in São Paulo, this must be one the priorities of the sugarcane sector.

4.7.4   Use of Fertilizers and Pesticides

In Brazil, the use of fertilizers in cane production is fairly small compared to other crops,[21] but the overall use can be significant in regions where the cane production is large.[24] The use of mineral fertilizer is partially avoided by the use of nutrient rich industrial wastes, such as filter cake and vinasse. Vinasse is a by product from the distillation of the sugarcane juice and has a high organic matter and potassium content, but relatively poor nitrogen, calcium, phosphorus and magnesium content.[24] For each liter of ethanol produced, 10–12 liters of vinasse are produced.

Advantages and disadvantages of vinasse application are described in literature.[21] Advantages include the reduced need for mineral fertilizers and a rise in pH, increased cation exchange capacity, increased availability of nutrients, improved soil structure, increased water retention and the development of soil micro-flora and fauna. Disadvantages are the risk of salinization and nutrient leaching, although results obtained from tests so far indicate that there is no damaging impact on the soil or groundwater if application is not excessive.[24]

In Brazil the application of vinasse on the soil is known as the ferti-irrigation process and has been a common practice for about 30 years. So far no critical problem regarding contamination of the soil or ground water has been reported. The R&D effort aims at reducing the amount of vinasse produced per liter of ethanol. Other possibilities are the reduction of the disposed volume (through thermal concentration) or the reduction of the organic matter content through bio digestion.

Agrochemicals include herbicides, insecticides, fungicides, etc. In Brazil, the consumption of agrochemicals for sugar cane production is lower than the consumption in other crops, such citric, corn, coffee and soybean.[21] However, as well as fertilizers, the total amount of agrochemicals used in some areas can be substantial, due to the concentration of sugarcane plantations in some regions.

As long as more resistant sugarcane varieties are developed, the use of agrochemicals can be reduced. This is a general tendency in Brazil, also because of the widespread practice of biological control. A better option is the production of organic cane, with use of agrochemicals. Some Brazilian mills have worked with organic cane in order to produce special sugars; a 10 % increase in productivity after some years of practice has been reported.

4.7.5   Impacts on Local Economy and on Living Conditions

A comparative analysis was performed between regions with sugarcane production in the state of São Paulo and regions in the same state with no sugarcane activity. The analysis is based on 2000 data published by the United Nations Development Program (UNDP).[33] Identification of municipalities with sugarcane production in 2000 was based on.[34] According to data available, in 2000 sugarcane mills were located in 96 cities out of 645 cities of state of São Paulo. The population of the municipalities with sugarcane activity varied from 2000 to 500,000 inhabitants, and through the comparison just cities with the same population in range were considered. The main results are presented in Table 4.4.

Table 4.4 *Comparison between municipalities with and without sugar and ethanol production in state of São Paulo, 2000*

Parameter	Cities with sugarcane activity	Cities with no sugarcane activity
Number of municipalities	96	499
Population range (1000)	2.4 – 500	2.4 – 500
Average income (R$ 2000) x 1000	17,193	12,441
Income/habitant (R$ 2000)[1]	308.7 ± 72.7	272.7 ± 85.2
Average Gini index[2]	0.519	0.528
Share of total income of 20 % poorest[1]	3.97 % ± 0.84 %	3.61 % ± 1.04 %
Share of households with electricity supply	99,6 %	98,8 %
Human development index (HDI)[1]	0.80 ± 0.03	0.78 ± 0.03

Notes: [1] Average values ± standard deviation.
  [2] The Gini index is a measure of statistical dispersion and is commonly used as a measure of inequality of income distribution. The index varies from 0 to 1, being 0 equivalent to perfectly equality and 1 to a hypothetical situation in which just one person has all income.

It can be seen from Table 4.4. that, on average, sugarcane industry brings benefits to the economy in cities where sugar and ethanol production occurs. As the analysis is based on 2000 figures when sugarcane production was lower than now, current results should be better. Despite better results regarding total income and income per capita, results presented in Table 4.4 also show that wealth distribution is not significantly improved due to the production of sugar and ethanol, as the Gini index is basically the same for both groups of municipalities. In 2000 the average Gini index for Brazil was 0.561 and 0.525 for the state of São Paulo;[33] Brazil has one of the highest Gini coefficients in the world. The Gini coefficient of the sugarcane and ethanol production sector is lower compared to the national average and compared to various other economic sectors.[21]

Also from Table 4.4 it can be observed that living conditions are slightly better for the set of municipalities in which sugarcane industry exist, a conclusion based on better results for the share of households with access to electricity supply and on the Human Development Index (HDI). As a matter of comparison, it should be noticed that theses indexes correspond to, respectively, 86.6 % and 0.699 in Brazil, and 98.9 % and 0.779 in the state of São Paulo.

The production of sugarcane/ethanol is an important source of employment in Brazil, both directly and indirectly related with the activity. Formal direct jobs were estimated as about 750 thousand in 2002, being almost 50 % related to sugarcane production.[21] The current number of total jobs (permanent + temporary) is estimated as 1.3 million, being about 500,000 in the state of São Paulo. As of 2003, employees in the formal sector (those who possess formal working papers) were 59 %, being 85 % in São Paulo.[21,24] This figure is significantly larger than the figure for the whole Brazilian economy.

Another important figure to mention is that the average wages received by workers in the sugarcane production in late 1990s, in the state of São Paulo, were 80 % higher than those received by workers involved with other crops. Workers in the sugarcane production received at that time wages 40–50 % higher than workers in the service and in the industry sector, as long as the comparison is done for people with the same skills.[21] Despite these positive aspects, wages levels may be insufficiently high to prevent poverty,[24] as can be seen from the indicators presented in Table 4.4.

The main problem regarding sugarcane production is concerned with the working conditions during manual cane harvesting. Most of the direct jobs are for harvesting; the workers are migrants who move to the areas of plantations due to lack of economic alternatives in their native regions. At least in the state of São Paulo the tendency is to reduce the number of workers in this very tough activity, as mechanical harvesting will be fully implemented up to 2017. A single harvester can displace about 80 cane cutters and just some workers could be relocated to other functions. Thus, in order to avoid heavy unemployment, it is important to displace workers for machines during the period of the expansion of the ethanol industry.

4.7.6 Answering the Question about Sustainability

During more than 30 years of large-scale production of ethanol in Brazil results achieved regarding sustainability are remarkable, but it is clear that improvements can and must be reached. Considering the main principles of sustainable production described at the beginning of this section, the main advantage of Brazilian production is regarding reductions of GHG emissions. Results are well above the minimum saves European countries tend to ask (30 % regarding the emissions of a gasoline vehicle), but can even be improved with, for instance, phasing-out of sugarcane burning in the field, enlargement of electricity production and reduction of diesel oil consumption in the agriculture.

The most common questions rose about ethanol production in Brazil with concern to biodiversity protection in general and destruction of the Amazon forest in particular, besides concerns about working conditions during manual harvesting. Both issues should be properly addressed by Brazilian society despite the target of exporting ethanol.

Brazil has so far presented modest results avoiding the deforestation of rain forests (e.g. according to the Brazilian Ministry for the Environment,[35] the deforestation of the Amazon forest was reduced from 2.7 Mha/year to 1.0 Mha/year in four years), but it has not yet been proved that sugarcane expansion is directly or indirectly responsible for deforestation. Plantations of soybeans and pastures are pushing the borders of the Amazon forest and environmentalists blame sugarcane expansion for inducing this process. However, the issue is not to prove if this argument is right or wrong, but to stop deforestation as fast as possible. And even the sugarcane production in the state of São Paulo, thousands of kilometers far from the Amazon, should be improved in order to reduce the effects of monoculture and to protect biodiversity.

Regarding the living conditions of workers, it is not reasonable to conceive the preservation of manual harvesting as a way to keep the jobs of hundred of thousands of people. A modern country needs to create better job opportunities and to improve life quality for all citizens, and mainly the poorest. In addition, modernization of ethanol production is a vital condition to make it a real commodity.

Ethanol production in Brazil is heterogeneous from the point of view of sustainability. Part of the production is in condition to fulfill most of the principles under discussion in Europe, but part of the production still presents serious problems. In a general sense, production in the state of São Paulo is more close to fulfilling the sustainability standards that can be imposed by ethanol importers in the short run.[24] In addition, it is almost a consensus that law enforcement would be enough to assure sustainable production of ethanol in Brazil, and that it is necessary to take action on this.

4.8   Is the Brazilian Experience Replicable?

It would be pretentious to identify the Brazilian experience in large-scale ethanol production as a model for new producer countries, as many mistakes were made in the past and there are still problems to be solved. However, the Brazilian experience is clearly a very good case study.

A very important point to start with is to remember that Brazil has very favorable conditions for large-scale production of biofuels, considering land availability, weather conditions, labor force, knowhow and R&D capacity. Few countries in the world have similar conditions and, thus, new producer countries should have different targets regarding Brazil.

Currently it is quite normal to present the Brazilian experience as a success, but until late 1990s there were doubts about the future production of ethanol on a large scale. At that time it was believed that the best option was to phase out the production of hydrate ethanol, just keeping the production of anhydrous ethanol. The turning point was the launch of FFVs in 2003, when the costs of ethanol production were low. Low costs were achieved due to continuous technology development, but full deregulation of the sugarcane industry (including sugar and ethanol markets), as well as deregulation of the fuel markets, were vital to induce competition among ethanol producers.

Currently, FFVs are the perfect solution, taking into account the size of the ethanol market in Brazil, but FFVs are not an option for a new producer country. Due to the lack of a required infrastructure, ethanol should be introduced through fuel blends, and for a new producer country the ethanol share should be first defined as a function of the local capacity of production.

A new producer country should primarily focus on its own market, but wide trade opportunities will arise in the years to come. Sustainable production and feasible supply should be the main requirements of the main importer countries, and production chain should be organized from the very beginning to match these requirements. However, despite the opportunities on biofuels trade, careful decisions are required as biofuels markets in developed countries shall continue to be protected and few countries can be really competitive considering the prices and conditions that need to be imposed by consumers (e.g. quality standards and certified sustainable production). Thus, the decision to be an exporter country still requires more coordination, including diplomatic and policy actions at the international level.

In order to face the chicken and egg problem, previously mentioned, coordination is required and all main actions should be planned. Biofuels production should be organized and managed like all energy sectors but, as opposed to the electric and oil industries, many players are required in the biomass market. In this sense governments should play an essential role, especially in developing countries. It was relatively easy to coordinate the actions in case of Brazil because large-scale production of ethanol started during a military regime.

Another important issue is that subsidies are essential and will be required for many years, but it is fundamental to clearly impose conditions (e.g. improvements on productivity, cost reductions) and to define the time extent of the given benefits. This was not done appropriately in Brazil for many years and the results were an installed capacity larger than necessary in the short term and the entrance of producers not exactly interested in the future of the ethanol market.

Continuous technological development is a challenge that needs to be faced by an energy sector in general, and by biofuels producers in particular. Taking the example of Brazil regarding the continuous development of cane varieties, this has been essential also because of the need to avoid plagues and diseases that could devastate the production in few years. Some required developments are time consuming and must be oriented to specific targets, and these challenges are difficult to be overcome by a single new producer country. Cooperation between producer countries and between developed and developing countries should be explored. In this sense, Brazil has great expertise to share regarding ethanol production from sugarcane.

In addition, it seems naïve to imagine that a biofuel program would be able to induce by itself reforms that could solve structural problems in a certain country, such as wealth distribution and endemic poverty. However, a biofuel program should be planned and developed to be part of the evolutionary process that society needs. Brazilian production of ethanol is based on plantations, and few entrepreneurs have the ownership of farms and industries. There are thousands of sugarcane suppliers in Brazil (about 60,000, responsible for about 30–35 % of the sugarcane produced[36]), but in a given region the market is monopsonic, or close to that. The production based on cooperatives of producers is a challenge that needs to be faced by a new producer country.

Finally, as stated by Goldemberg,[37] large-scale production of ethanol from sugarcane can be replicated in other countries without serious damage to natural ecosystems. In order to displace about 10 % of the gasoline used in the world it will be necessary to expand sugarcane production by a factor of ten. The area required, about 30 Mha of sugarcane in Brazil and in other countries, mostly developing countries, is a small fraction of the more than 1 billion hectares of primary crops already harvested on the planet.

4.9   Conclusions

Large-scale production of ethanol in Brazil is a relevant case study, both because of the good results achieved and the mistakes done. The importance of ethanol as automotive fuel in Brazil is unique, and barely will be observed in another country. Brazil is so far the unique country where ethanol is competitive with gasoline with no subsidies and where there is real competition in the market between these two fuels.

Ethanol production costs in Brazil must be the lowest in the world for many years and perhaps ever. However, in order to keep its position as a leading country and – mainly – to be really competitive in the international market, it is necessary to maintain R&D efforts, while considering different routes of production (e.g. hydrolysis and biofuels production based on biomass gasification) and different fuels (e.g. the production of diesel oil through synthesis gases). In this sense, diversification of production is an essential issue and it is necessary to go much further than the combined production of fuels and electricity, and go for chemicals, raw materials, lubricants, etc.

A crucial point for Brazil is to enhance sustainability of biofuels production. This challenge should be faced not just because of the opportunities for trading, but mainly because this is the only way to go. For most of the improvements required, law enforcement, investments and the use of adequate technology will be enough. However, the real challenge in Brazil is to improve wealth distribution and to prevent poverty. This is not a target to be faced only by the ethanol industry, but a good example should be given by one of the most successful economic sectors in the country.

Brazil can produce and export a reasonable amount of biofuels but must, above all, exchange its expertise with other countries in order to make ethanol a real commodity.

References

1. Unica – União da Indústria de Cana de Açúcar (2007) Available at www.unica.com.br
2. RFA – Renewable Fuels Association. Statistics data (2006) Available at http://www.ethanolrfa.org/industry/statistics/#C
3. F. Rosillo-Calle, A. Walter (2006) Global market for bioethanol: historical trends and future prospects, *Energy for Sustainable Development* **10**(1): 20–32.
4. J. Goldemberg, S.T. Coelho, P.M. Nastari, O. Lucon (2004) Ethanol learning curve – the Brazilian experience, *Biomass and Bioenergy* **26**(3): 301–4.
5. S. Coelho, A. Walter, J. Goldemberg, D. Schiozer, F. Moreira et al. (2006) Efficiencies and Infrastructure Brazil – a country profile on sustainable development, Chapter 4 of *Indigenous Energy Technologies*, 1st ed. International Atomic Energy Agency, Vienna, p. 252.
6. ANP – Brazilian Oil Agency (2007) Data available at www.anp.gov.br
7. CENEA – USP. Center of Applied Economics on Agriculture – University of São Paulo (2007) Data available at www.cepea.esalq.usp.br/indicador/alcool/
8. Brazil. Ministry of Mines and Energy (MME) (2006) *Brazilian Energy Balance 2006*, MME Press, Brasília, p. 192. Available at www.mme.gov.br/site/menu/
9. Unica – União da Indústria de Cana de Açúcar (2006) Presentation at VI International Conference Datagro, São Paulo. Available at www.unica.com.br
10. EPE – Empresa de Pesquisa Energética (2006) National Energy Plan 2030 – *strategy for enlarging energy supply*. Presentation at the Ministry of Mines and Energy, Brasília. Available at www.epe.gov.br
11. FAO – Food and Agriculture Organization (2007) *Statistics Data*. Available at www.faostat.fao.org
12. J. Romm (2006) The car and fuel of the future, *Energy Policy* **34**: 2609–14.
13. J.D. van den Wall Bake (2006) *Cane as key in Brazilian ethanol industry – Understanding cost reductions through an experience curve approach*, Masters thesis, Utrecht University, Utrecht, p. 82.
14. A. Walter (2006) Is Brazilian biofuels experience a model for other developing countries, *Entwicklung & Ländlicher Raum* **6**: 22–4.
15. F.O. Lichts (2006) World Ethanol & Biofuels Report, **5**(7).
16. M.C.B. Cavalcanti (2006) *Análise dos Tributos Incidentes sobre os Combustíveis Automotivos no Brasil*, Masters thesis, Federal University of Rio de Janeiro, Rio de Janeiro, p. 213.
17. L.C.C. Carvalho (2001) Hora da virada, *Agroanalysis* **21**(9).

18. FIPE – Fundação Instituto de Pesquisas e Tecnológicas (2001) *Scenarios for the sugar and alcohol sector*, Fundação Instituto de Pesquisas e Tecnológicas, São Paulo University (in Portuguese).
19. O. Henniges, J. Zeddies (2004) Fuel Ethanol Production in USA and Germany – a cost comparison, F.O. Lichts World Ethanol and Biofuels Report, 1(11).
20. Worldwatch Institute (2006) *Global Potential and Implications for Sustainable Agriculture and Energy in the 21st Century*, Washington, p. 417.
21. I.C. Macedo (2005) *Sugar cane's energy – Twelve studies on Brazilian sugar cane agribusiness and its sustainability* (ed: I.C. Macedo), Unica, São Paulo, Brazil, p. 237.
22. O. Braunbeck, A. Bauen, F. Rosillo-Calle, L. Cortez (1999) Prospects for green cane harvesting and cane residue use in Brazil, *Biomass and Bioenergy*, 495–506.
23. J. Finguerut (2007) Optimization of Bioethanol Production. Presentation at the LA-EU Biofuels Research Workshop, Campinas, Brazil.
24. E. Smeets, M. Junginger, A. Faaij, A. Walter, P. Dolzan (2006) *Sustainability of Brazilian bio-ethanol*, Report NWS-E-2006-110, University of Utrecht.
25. O. Braunbeck, L. Cortez (2000) *Industrial Uses of Biomass Energy – The Example of Brazil* (ed: F. Rosillo-Calle, S.V. Bajay, H. Rothman), Taylor & Francis, London, Chapter 5.
26. I. Macedo, L. Cortez (2000) *Industrial Uses of Biomass Energy – The Example of Brazil* (ed: F. Rosillo-Calle, S.V. Bajay, H. Rothman), Taylor & Francis, London, Chapter 6.
27. J.R. Moreira, J. Goldemberg (1999) 'The alcohol program', *Energy Policy* **27**(4): 229–45.
28. J. Cramer et al. (2007) Testing Framework for Sustainable Biomass – Final report from the project group 'Sustainable Production of Biomass', Amsterdam, 72.
29. I.C. Macedo, M.R.L.V. Leal, J.E.A.R. da Silva (2004) *Assessment of greenhouse gas emissions in the production and use of fuel ethanol in Brazil*. Accessible via: http://www.unica.com .br/i_pages/files/pdf_ingles.pdf, Secretariat of the Environment of the State of São Paulo, Brazil.
30. IEA – International Energy Agency (2004) *Biofuels for Transport: An International Perspective*, IEA/OECD, Paris.
31. USDA – US Department of Agriculture (2006) *The Economics of Bioethanol Production in the EU 2006*, USDA Foreign Agricultural Service, Grain Report E36081.
32. S.T. Coelho (2005) Biofuels – advantages and trade barriers. In: Expert Meeting on the Developing Countries Participation in New and Dynamic Sectors of World Trade. Report by UNCTAD/DITC/TED/2005/1, Geneva.
33. UNDP – United Nations Development Program (2004) Atlas do desenvolvimento humano do Brasil. UNDP, Brasília. Available at http://www.pnud.org.br/atlas/tabelas/index.php
34. Unica – União da Indústria de Cana de Açúcar (2007) Informação Única, 39. Available at www.unica.com.br
35. Brazil. Ministry for the Environment (MMA). Information available at www.mma.gov.br
36. I.C. Macedo, L.A.H. Nogueira (2005) *Biocombustíveis*, Núcleo de Assuntos Estratégico da Presidência da República, Brasília, Brazil, p. 235.
37. J. Goldemberg (2007) Ethanol for a Sustainable Energy Future, *Science* **315**, 808–10.

5

Process Technologies for Biodiesel Production

Martin Mittelbach

Department of Renewable Resources, Institute of Chemistry,
Karl-Franzens-University, Graz, Austria

5.1 Introduction

In the last years bio energy and biofuels have become a major issue around the world, especially because of the rising oil prices on the one hand, and of public awareness and concern on energy safety and environmental issues like climate change on the other hand. In particular, fatty acid methyl esters, also called 'biodiesel', have become the leading biofuel for diesel engines, because they have very similar fuel properties to fossil fuel and can be used without any changes to the engine. So the acceptance of engine manufacturers and the mineral oil industry is very high. Another reason for the quick market penetration of biodiesel is the relatively easy process technology, which can be done on a small as well as industrial scale. This chapter will give an overview of the current situation of biodiesel production worldwide and also of possible feedstocks. In particular the chemical principles of biodiesel production as well as the different production technologies will be highlighted.

5.2 Biodiesel Production Worldwide

Europe has been the leading region for biodiesel production. The first pilot plant for the production of biodiesel from rape seed oil has been installed in Austria in 1987 in

Table 5.1   *Biodiesel production 2006 in Europe in 1000 tons*[3]

Germany	2,662
France	743
Italy	447
UK	192
Austria	123
Poland	116
Czech Rep.	107
Spain	99
Portugal	91
Slovakia	82
Denmark	80
Greece	42
Belgium	25
The Netherlands	18
Sweden	13
Slovenia	11
Romania	10
Lithuania	10
Latvia	7
Bulgaria	4
Ireland	4
Malta	2
Cyprus	1
Estonia	1
Finland	0
Hungary	0
Luxemburg	0
TOTAL	**4,890**

Silberberg, resulting from several years of research at the Institute of Chemistry of the University of Graz.[1] The possibility of using pure biodiesel without paying mineral tax has led to the installation of several industrial scale plants in 1991 in Austria and Germany, but very soon countries like Italy, France and the Czech Republic followed. The period of slow but constant development of biodiesel activities in Europe for several years ended with the European directive for the promotion of biodiesel in the year 2003,[2] demanding a market share of biodiesel of 5.75 % for the transport fuel in 2010. Since then the development of biodiesel production almost exploded in all 27 countries of the European Union, leading to a total production of biodiesel of 5.6 million tons in the year 2006 and a production capacity of almost 10 million tons. An overview of the European biodiesel production is given in Table 5.1.

The production has been increased by approx. 50 % from the year 2005 until 2006; the production capacity in 2007 will exceed 10 million tons per year.[3] The almost exploding development of biodiesel production in Europe will be cooled down by the lack of feedstocks and raw material, which partly even today has to be imported.

Almost in every country worldwide biodiesel activities have been initiated; the leading countries are also those with the largest production of vegetable oils. In 2006 in the US the production of biodiesel exceeded 1 million tons per year; however, biodiesel plants have been installed with an overall production capacity of over 5 million tons per year.[4] But also the major palm oil producing countries like Malaysia and Indonesia have a series of biodiesel plants already installed with a capacity of over 1 million tons per year. Similar activities can be found in China and India with their huge demand for transport fuel, but also in the vegetable oil producing countries like Brazil and Argentina.

5.3   Feedstocks for Biodiesel Production

Today a production capacity of almost 30 million tons of biodiesel exists worldwide. On the other hand there is a total annual production of vegetable oils of approx. 110 million tons per year, which is mainly used for food purposes. As the production of vegetable oils cannot be increased in such a way as there is the demand for biodiesel, competition with the food market will be inevitable. Also concern on unsustainable production of oil plants like palm oil has led to extensive discussions leading to the search for non-edible oil seeds.

Basically all vegetable oils and animal fat can be used as feedstock for biodiesel production. Most of these oils and fats have a similar chemical composition, they consist of triglycerides with different amounts of individual fatty acids. The major fatty acids are those with a chain length of 16 and 18 carbons, whereas the chain could be saturated or unsaturated. Methyl esters produced from these fatty acids have very similar combustion characteristics in a diesel engine, because the major components in fossil diesel fuel are also straight chain hydrocarbons with a chain length of about 16 carbons (hexadecane, 'cetane'). The major differences between the methyl esters from different feedstocks refer to the amount of unsaturated fatty acids. The best combustion characteristics as well as oxidation stability come from saturated fatty acids; however, cold temperature behaviour is worse due to the high melting points of these fatty acids. On the other hand, high unsaturation of fatty acids leads to optimum low temperature properties, but the oxidation and storage stability is worse.

The major feedstocks for the biodiesel production today are rape seed oil (Canola), soybean oil and palm oil. The fuel properties of the methyl esters out of these oils are quite similar except for the poor cold temperature behaviour of palm oil because of the high portion of saturated fatty acids. However, depending on the climate conditions of a country, an optimum mix of methyl esters out of these feedstocks can be used. Also a series of other vegetable oils has quite similar fatty acid distribution and can also be used as blend. Only coconut oil and palm kernel oil have fatty acids with 12 or 14 carbons as major components. Therefore the methyl esters out of these fats have lower boiling points, but could be used perfectly as ad-mixture to common biodiesel.

Especially in Asian countries like India and China the use of non-edible seed oils for biofuel production is very popular; in that case there would be no competition with the food production, especially when these oil plants are grown on marginal areas not suitable for food production. Especially *Jatropha curcas* L. has attracted enormous attention within the last years, especially in India, Indonesia and in the Philippines. As there would be no competition with the food production and also with the traditional agricultural areas Jatropha could fill the gap between actual vegetable oil production and demand for biofuels.

Another interesting feedstock for biodiesel production is oil produced from algae, which could be grown in open ponds or in closed tubes. The productivity is estimated to be much higher per area than with traditional oil seeds and furthermore there is no need for agricultural land, it can be produced at any place, where water and sunlight are existing.

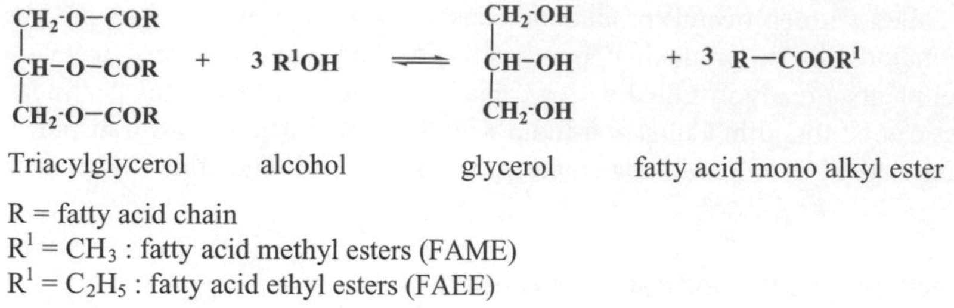

$$
\begin{array}{c}
\text{CH}_2\text{-O}-\text{COR} \\
| \\
\text{CH}-\text{O}-\text{COR} \\
| \\
\text{CH}_2\text{-O}-\text{COR}
\end{array}
\quad + \quad 3\ \text{R}^1\text{OH}
\quad \rightleftharpoons \quad
\begin{array}{c}
\text{CH}_2\text{-OH} \\
| \\
\text{CH}-\text{OH} \\
| \\
\text{CH}_2\text{-OH}
\end{array}
\quad + \quad 3\ \text{R}-\text{COOR}^1
$$

Triacylglycerol   alcohol   glycerol   fatty acid mono alkyl ester

R = fatty acid chain
$R^1 = CH_3$: fatty acid methyl esters (FAME)
$R^1 = C_2H_5$: fatty acid ethyl esters (FAEE)

Figure 5.1 *Production of fatty acid mono alkyl esters via transesterification.*

However, today an economic production of biodiesel from algae does not seem to be very realistic, but further research in that area will be necessary.

5.4   Chemical Principles of Biodiesel Production[5]

Fatty acid methyl esters have been known for over 150 years. The first description of the preparation of the esters was published in 1852.[6] However, for a long time fatty acid methyl esters were mainly used as derivatives for analyzing the fatty acid distribution of fats and oils, so the preparation mainly was done in analytical scale. Since the mid 20th century fatty acid methyl esters have become a major oleo chemical commodity as intermediate for the production of fatty alcohols, used for the production of non-ionic detergents. But only since the late Seventies have fatty acid methyl esters have been tested and used as diesel fuel substitute.

Chemically biodiesel is equivalent to fatty acid methyl esters or ethyl esters, produced out of triacylglycerols via transesterification or out of fatty acids via esterification. In Figure 5.1 the formula scheme for the production of fatty acid mono alkyl esters out of triacylglycerol is shown. Fatty acid methyl esters today are the most commonly used biodiesel species, whereas fatty acid ethyl esters (FAEE) so far have been only produced in laboratory or pilot scale.

Figure 5.2 shows the chemical equation for an esterification reaction. As vegetable oils or animal fats mainly consist of triacylglycerol (triglycerides) the main reaction for the production of biodiesel is the transesterification or alcoholysis reaction, whereas esterification is only necessary for feedstocks with higher content of free fatty acids.

In a transesterification or alcoholysis reaction one mole of triglyceride reacts with three moles of alcohol to form one mole of glycerol and three moles of the respective fatty acid alkyl ester. The process is a sequence of three reversible reactions, in which the triglyceride molecule is converted step by step into diglyceride, monoglyceride and glycerol. In order to shift the equilibrium to the right, methanol is added in an excess over the stoichiometric amount in most commercial biodiesel production plants. A main advantage of methanolysis

$$
\text{R-COOH} \ + \ \text{R}^1\text{OH} \quad \rightleftharpoons \quad \text{R-COOR}^1 \ + \ \text{H}_2\text{O}
$$

Fatty acid   alcohol   fatty acid mono alkyl ester   water

R = fatty acid chain

Figure 5.2 *Production of fatty acid mono alkyl esters via esterification.*

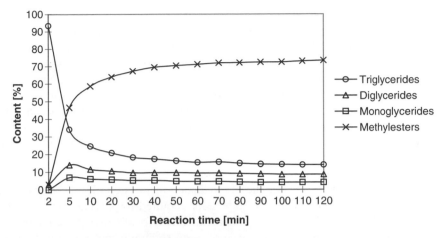

Figure 5.3 *Schematic course of a methanolysis reaction.*
Reaction conditions: sunflower oil: methanol = 3:1 (mol/mol), 0.5 % KOH, T = 25 °C
Source: adapted from [8], reproduced by permission of Wiley VCH.

as compared to transesterification with higher alcohols is the fact that the two main products, glycerol and fatty acid methyl esters (FAME), are hardly miscible and thus form separate phases – an upper ester phase and a lower glycerol phase. This process removes glycerol from the reaction mixture and enables high conversion. Ester yields can even be increased – while at the same time minimizing the excess amount of methanol – by conducting methanolysis in two or three steps. Here only a portion of the total alcohol volume required is added in each step, and the glycerol phase produced is separated after each process stage.[7] Finally, regardless of the type of alcohol used, some form of catalyst has to be present to achieve high ester yields under comparatively mild reaction conditions.

Figure 5.3[8] illustrates the schematic course of a typical methanolysis reaction. Whereas the concentration of triglycerides as the starting material decreases and the amount of methyl esters as the desired product increases throughout the reaction, the concentrations of partial glycerides (i.e. mono- and diglycerides) reach a passing maximum.

The esterification reaction according to Figure 5.2 is a typical equilibrium reaction, so to increase the yield of fatty acid alkyl esters it is necessary to use an excess of alcohol or to remove one of the end products out of the equilibrium, e.g. the water by distillation or by the use of concentrated sulphuric acid.

In order to increase the reaction rate of transesterification or esterification in most cases catalysts are used.

5.5 Catalysts for Transesterification and Esterification Reactions

5.5.1 Alkaline Catalysis

Alkaline or basic compounds are by far the most commonly used catalysts for biodiesel production. The main advantage of this form of catalysis over acid-catalyzed transesterifications is the high conversion rate under mild conditions in comparatively short reaction times.[9] So it was estimated that under the same temperature conditions and catalyst concentrations methanolysis might proceed about 4000 times faster in the presence of an alkaline catalyst than in the presence of the same amount of an acidic equivalent.[10] Moreover, alkaline catalysts are less corrosive to industrial equipment, so that they enable the use of less expensive carbon-steel reactor material. The main drawback of the technology is the sensitivity of basic catalysts to free fatty acids contained in the feedstock material. This means that alkali-catalyzed transesterifications optimally work with high-quality, low-acidic vegetable oils, which are however more expensive than waste oils. If low-cost materials, such as waste fats with a high amount of free fatty acids, are to be processed by alkaline catalysis, deacidification or pre-esterification steps are required.

Today most of the commercial biodiesel production plants are utilizing homogeneous, alkaline catalysts. Traditionally the alkoxide anion required for the reaction is produced either by using directly sodium or potassium methoxide or by dissolving sodium or potassium hydroxide in methanol. The advantage of using sodium or potassium methoxide is the fact that no additional water is formed and therefore side reactions like saponification can be avoided. The use of the cheaper catalysts sodium or potassium hydroxide leads to the formation of methanolate and water, which can lead to increased amounts of soaps. However, because of the fact that glycerol separates during alcoholysis reactions, also water is removed out of the equilibrium, so under controlled reaction conditions, saponification can be kept to a minimum. By comparison of different alkaline catalysts for the methanolysis of sunflower oil reactions using sodium hydroxide turned out to be the fastest.[11]

The amount of alkaline catalyst depends on the quality of the oil, especially on the content of free fatty acids. Under alkaline catalysis free fatty acids are immediately converted into soaps, which can prevent the separation of glycerol and finally can lead to total saponification of all fatty acid material. So the alkaline catalysis is limited to feedstocks up to a content of approx. 3 % of fatty acids.

There are also other alkaline catalysts like guanidines or anion exchange resins described in literature, however, no commercial application in production plants is known. The catalysis with guanidine carbonate has the advantage that after phase separation of ester and glycerol phases during workup of the glycerol phase the catalyst decomposes during distillation of the glycerol into ammonia and carbon dioxide.[12] Ammonia could be trapped in phosphoric or sulphuric acid giving ammonium salts suitable as fertilizer. But in any case the homogenous catalysts cannot be reused. (See Table 5.2.)

Table 5.2   *Overview of homogenous alkaline catalysts*

Type of catalyst	Comments
Sodium hydroxide	Cheap, disposal of residual salts necessary
Potassium hydroxide	Reuse as fertilizer possible, fast reaction rate, better separation of glycerol
Sodium methoxide	No dissolution of catalyst necessary, disposal of salts necessary
Potassium methoxide	No dissolution of catalyst necessary, use as fertilizer possible, better separation of glycerol, high price
Guanidines	Higher price, purification of glycerol easier

5.5.2   Acid Catalysis

Acid catalysis offers the advantage of also esterifying free fatty acids contained in the fats and oils and is therefore especially suited for the transesterification of highly acidic fatty materials.

However, acid-catalyzed transesterifications are usually far slower than alkali-catalyzed reactions and require higher temperatures and pressures as well as higher amounts of alcohol. The typical reaction conditions for homogeneous acid-catalyzed methanolysis are temperatures of up to 100 °C and pressures of up to 5 bars in order keep the alcohol liquid.[13] A further disadvantage of acid catalysis – probably prompted by the higher reaction temperatures – is an increased formation of unwanted secondary products, such as dialkylethers or glycerol ethers.[14]

Because of the slow reaction rates and high temperatures needed for transesterification acid catalysts are only used for esterification reactions. So for vegetable oils or animal fats with an amount of free fatty acids over approx. 3 % two strategies are possible. The free fatty acids can either be removed by alkaline treatment or can be esterified under acidic conditions prior to the alkaline catalyzed transesterification reaction. This so-called pre-esterification has the advantage that prior to the trans-esterification most of the free fatty acids are already converted into FAME, so the overall yield is very high. If you have to remove the free fatty acids prior to the trans-esterification, similar to the deacidification of vegetable oils during refining, you don't have to change the transesterification conditions; however, these fatty acids are lost in the overall yield unless these fatty acid are esterified again in a separate step.

The cheapest and well-known catalyst for esterification reactions is concentrated sulphuric acid. The main disadvantages of this catalyst are the possibility of the formation of side products like dark colored oxidized or other decomposition products. As organic compound also p-toluene sulphonic acid can be used; however, the high price of the compound has prevented broader application. As heterogeneous catalyst also cationic ion exchange resins can be used in continuous reaction columns; however, this approach has only been used so far in pilot plants. The esterification of free fatty acids with methanol at increased temperatures above the boiling point of methanol at ambient pressure was achieved by introducing methanol into a preheated reaction mixture,[15] containing free fatty acids and an acid catalyst. The reaction rates were significantly higher than under reflux temperature. (See Table 5.3.)

5.5.3   Heterogeneous Catalysis

Whereas traditional homogeneous catalysis offers a series of advantages, its major disadvantage is the fact that homogenous catalysts cannot be reused. Moreover, catalyst residues

Table 5.3   *Overview of acidic catalysts*

Type of catalyst	Comments
Conc. Sulphuric acid	Cheap, decomposition products, corrosion
p-Toluene-sulphonic acid	High price, recycling necessary
Acidic ion exchange resins	High price, continuous reaction possible, low stability

Table 5.4   *Overview on heterogeneous catalysts (for references see [5])*

Catalyst type	Examples
Alkali metal carbonates and hydrogen carbonates	Na_2CO_3, $NaHCO_3$ K_2CO_3, $KHCO_3$
Alkali metal oxides	K_2O (produced by burning oil crop waste)
Alkali metal salts of carboxylic acids	Cs-laurate
Alkaline earth metal alcoholates	Mixtures of alkali/alkaline earth metal oxides and alcoholates
Alkaline earth metal carbonates	$CaCO_3$
Alkaline earth metal oxides	CaO, SrO, BaO
Alkaline earth metal hydroxides	$Ba(OH)_2$
Alkaline earth metal salts of carboxylic acids	Ca- and Ba-acetate
Strong anion exchange resins	Amberlyst A 26, A 27
Zink oxides/ aluminates	
Metal phosphates	ortho-phosphates of aluminum, gallium or iron (III)
Transition metal oxides, hydroxides and carbonates	Fe_2O_3 (+ Al_2O_3), Fe_2O_3, Fe_3O_4, FeOOH, NiO, Ni_2O_3, $NiCO_3$, $Ni(OH)_2$ Al_2O_3
Transition metal salts of amino acids	Zn- and Cd-arginate
Transition metal salts of fatty acids	Zn- and Mn-palmitates and stearates
Silicates and layered clay minerals	Na-/K-silicate Zn-, Ti- or Sn-silicates and aluminates
Zeolite catalysts	titanium-based zeolites, faujasites

have to be removed from the ester product, usually necessitating several washing steps, which increases production costs. Thus there have been various attempts at simplifying product purification by applying heterogeneous catalysts, which can be recovered by decantation or filtration or are alternatively used in a fixed-bed catalyst arrangement. The most frequently cited heterogeneous alkaline catalysts are alkali metal- and alkaline earth metal carbonates and oxides (see Table 5.4.) For the production of biofuels in tropical countries, Graille et al. recommended utilizing the ashes of oil crop waste (e.g. coconut fibres, shells and husks).[16] The resulting natural catalysts are rich in carbonates and potassium oxide and have shown considerable activity in transesterifications of coconut oil with methanol and water free ethanol.

Among the catalysts listed in Table 5.4, the application of calcium carbonate may seem particularly promising, as it is a readily available, low-cost substance. Moreover, the catalyst showed no decrease in activity even after several weeks of utilization.[17] However, the high reaction temperatures and pressures and the high alcohol volumes required in this technology are likely to prevent its commercial application. The reaction conditions described sometimes are so drastic, that there might be also conversion without any use of catalyst. Mostly, comparison experiments without any catalyst are missing in the experiments.

Similar drawbacks have to be attested for alkali metal or alkaline earth metal salts of carboxylic acids. The use of strong alkaline ion-exchange resins, on the other hand, is limited by their low stability at temperatures higher than 40 °C and by the fact that free fatty acids in the feedstock neutralize the catalysts even in low concentrations. Finally, glycerol released during the transesterification process has a strong affinity to polymeric resin material, which can result in complete impermeability of the catalysts.[18] Transesterification reactions with triolein and ethanol using various ion-exchange resins were conducted, showing that anion-exchange resin with a lower cross-linking density and a smaller particle size gave the highest reaction rates and conversions.[19] At the moment no commercial biodiesel production plant is operating on heterogeneous alkaline catalysis as the sole transesterification strategy.

Most recently, the first technology using heterogeneous catalysts like zink oxides or zinc aluminates, will be used in several commercial biodiesel production plants in France and in the USA.[20,21] The so-called Esterfip-H process was developed by the Institut Français du Pétrole (IFP) and is designed and commercialized by Axens. The main advantages of the process are described as the production of high-quality glycerol and no need for disposal of salts resulting from the catalyst. However, the overall economic advantages have to be proved in long-term running.

5.5.4   Enzymes as Catalysts

In addition to the inorganic or metallo-organic catalysts presented so far, also the use of lipases from various micro organisms has become a topic in biodiesel production. Lipases are enzymes which catalyze both the hydrolytic cleavage and the synthesis of ester bonds in glycerol esters. Their application in FAME production dates back to Choo and Ong,[22] filing a patent application on lipase-catalyzed methanolysis in the presence of water, and to Mittelbach, reporting on the first water-free process for lipase-catalyzed biodiesel production.[23] As compared to other catalyst types, biocatalysts have several advantages. They enable conversion under mild temperature-, pressure- and pH-conditions. Neither the ester product nor the glycerol phase has to be purified from basic catalyst residues or soaps. That means that phase separation is easier, high-quality glycerol can be sold as a by-product, and environmental problems due to alkaline wastewater are eliminated.[24] Moreover, both the transesterification of triglycerides and the esterification of free fatty acids occur in one process step. As a consequence, also highly acidic fatty materials, such as palm oil or waste oils, can be used without pre-treatment.[25] Finally, many lipases show considerable activity in catalyzing transesterifications with long or branched-chain alcohols, which can hardly be converted to fatty acid esters in the presence of conventional alkaline catalysts.

However, lipase-catalyzed transesterifications also entail a series of drawbacks. As compared to conventional alkaline catalysis, reaction efficiency tends to be poor, so that biocatalysts usually necessitates far longer reaction times and higher catalyst concentrations. The main hurdle to the application of lipases in industrial biodiesel production is their high price, especially if they are used in the form of highly purified, extra cellular enzyme preparations, which cannot be recovered from the reaction products. One strategy to overcome this difficulty is the immobilization of lipases on a carrier, so that the enzymes can be removed from the reaction mixture and can theoretically be reused for subsequent transesterifications. Immobilization also offers the advantage that in many cases the fixed lipases tend to be more active and stable than free enzymes. Traditional carrier materials

Table 5.5 *Critical conditions of different solvents*

	T_{crit} [°C]	P_{crit} [MPa]	ρ_{crit} [kg m^{-3}]
Water (H_2O)	373.9	22.06	322
Carbon dioxide (CO_2)	30.9	7.375	468
Methanol (CH_3OH)	239.4	8.092	272
Ethanol (CH_3CH_2OH)	240.7	6.137	276
1-Propanol ($CH_3CH_2CH_2OH$)	263.6	5.170	275

(such as anion exchange resins or polyethylene) can be replaced by renewable, readily available substances like corn cob granulate.[26]

5.6 Transesterification in Supercritical Alcohols

Basically, transesterification of triglycerides with lower alcohols also proceeds in the absence of a catalyst, provided reaction temperatures and pressures are high enough.[27,28] Above the critical temperature a gas cannot be liquefied. The supercritical conditions show liquid as well as gaseous properties, so there have been a lot of applications for CO_2 extractions but also for carrying out chemical reactions. An overview of biodiesel production using supercritical methods is given in [29]. In Table 5.5 the critical conditions for different alcohols are outlined and Figure 5.4 gives the conversion rates of alcoholysis with different alcohols at 300 °C. The advantages of not using a catalyst for transesterification are that high-purity esters and soap-free glycerol are produced (see also Table 5.6). The high excess of methanol which has to be used during supercritical transesterification seems to make the process not economically feasible, however, a two-step process has been described, which in the first step hydrolyzes the glycerides into fatty acid with an excess of water, and in the second step esterification takes place, which requires lower amounts of methanol.[30,31]

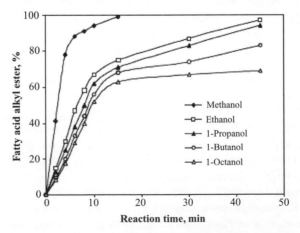

Figure 5.4 *Conversion rates during alcoholysis under supercritical condition.*[29]

Table 5.6   *Comparison between classical and supercritical methanolysis*

	Base catalyzed methanolysis	SCM method
Catalyst	Alkali hydroxides, alcoholates	none
Methanol amount	Slight excess	High excess
Reaction temperature [°C]	20–60	250–300
Reaction pressure [MPa]	0.1	10–25
Reaction time	30–120	7–15
Free fatty acids	soaps	FAME, water
Purification of glycerol	Salt formation	No salts, possible condensation products (methyl ethers)
Energy consumption	low	high

5.7   Alternative Approaches

Classical alkaline catalyzed transesterification reactions are multiphase reactions. In the beginning methanol is not soluble in the vegetable oil, but during increased formation of fatty acid methyl esters the reaction mixture becomes homogenous until the formation of glycerol begins, which again is insoluble in the methyl ester phase and separates at the bottom. This fact facilitates the completeness of the reaction by separation of the end product out of the reaction mixture. However, also the catalyst is removed together with the glycerol. In order to overcome the mass transfer limitations the use of the co-solvent Tetrahydrofuran (THF, Oxolane) was suggested.[32] However, for a technical application the solvent has to be evaporated after the reaction which consumes quite a lot of energy.

A novel biodiesel-like material was developed by reacting soybean oil with dimethyl carbonate, which avoided the co-production of glycerol.[33] The main difference between this new route and classical biodiesel production is the presence of fatty acid glycerol carbonate monoesters in addition to FAME. The presence of these compounds influence both fuel and flow properties. Also the microwave assisted alcoholysis has been reported, however, the evidence of significant improvement over classical routes is still missing, also an industrial scale application does not seem to be realistic.[34]

For complete conversion of oil directly from oil-containing seeds or other materials the so-called in-situ transesterification can be used. That means that the lower alcohol serves both as an extracting agent for the oils and the reagent for alcoholysis. In-situ transesterification offers a series of advantages. First, hexane is no longer necessary as a solvent in oil recovery. Second, the whole oil seed is subjected to the transesterification process, so that losses due to incomplete oil production are minimized. Finally, the esterified lipids tend to be easier to recover from the solid residue than native oils due to their decreased viscosities.[35] The in situ production of FAME with the use of supercritical fluids or microwaves also has been suggested.[36,37]

5.8   Overview of Process Technologies

In the oleo chemical industry the production of fatty acid methyl esters has a long tradition, because these products are an important intermediate for the further production of fatty alcohols and fatty alcohol ethoxylates. These production units mainly use sodium methoxide as catalyst under more drastic reaction conditions like higher pressure and high temperatures. Under these reaction conditions also free fatty acids are converted into fatty acid methyl esters. For the use as intermediates for fatty alcohol production the esters have to be distilled. At the beginning of the biodiesel development these technologies, however, were too expensive and needed too high investment costs, Therefore the so-called low temperature and low pressure processes were developed, which use temperatures up to 60 °C at ambient pressure and there is no need for distillation of the final product. These processes can be used in very small production units, but today also production plants with a capacity of 100,000 tons or more are using this technology. This is also the reason that today there is an enormous number of small-scale producers existing worldwide, because one can use simple equipment. However, these so-called 'backstage' or 'garage' producers mostly don't have any safety precautions or quality control of the product, furthermore they cannot further process the glycerol layer, which contains excess methanol and is therefore considered as special waste.

Moreover, in most cases, the biodiesel produced does not meet the high quality standards defined in the European specifications EN 14214 or ASTM specifications D 6584, and therefore it is possible that this quality can lead to serious injection pump or engine problems.

Therefore in the following only the technologies with industrial applications are described.

There are different possibilities in classifying the different biodiesel production technologies. One can distinguish according to the type of catalyst between homogenously or heterogeneously catalyzed processes; one can distinguish according to the reaction conditions between low and high temperature and pressure reactions, or between continuous or batch operation. On the other hand it also possible to classify according to type of feedstocks. The so-called single feedstock technologies are using half or fully refined vegetable oils like rape seed, soybean, sunflower etc. With these technologies the content of free fatty acids should be very low, so the formation of soaps is very limited. Normally alkaline catalysts like sodium methoxide or potassium hydroxide is used, and the soaps formed as side products during the reaction are either removed by water washing steps or recycled by esterification with acid catalysts after work up of the glycerol phase. With this technology also a small amount of other feedstocks like recycled frying oil or higher acidic palm oil can be blended to the refined vegetable oils.

The so-called multi-feedstock technologies are also capable of processing feedstocks with higher amounts of free fatty acids. Here a so-called preesterification of the free fatty acids is necessary, or during a high pressure and temperature process all fatty material is directly converted in FAME in one step. These processes could be capable of processing any type of feedstocks, including acid oils, animal fat, high acidic palm oil or even fatty acids. The reaction conditions can be easily adapted to the change of feedstock.

Though a differentiation of these two technologies is often not very easy, especially with newer developments of technology, the terms single and feedstock technologies are broadly used in the biodiesel terminology, and therefore will be used in the following.

5.8.1   Single Feedstock Technologies

The biggest biodiesel production units with a capacity of over 100,000 tons mainly use fully refined vegetable oils with low content of water and free fatty acids. In that case

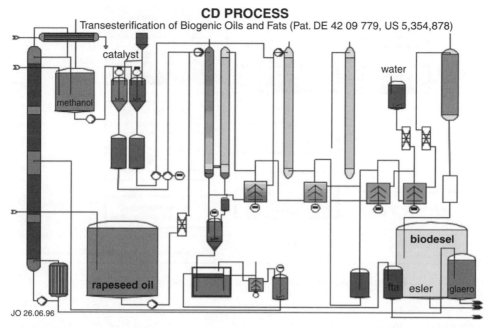

CD PROCESS
Transesterification of Biogenic Oils and Fats (Pat. DE 42 09 779, US 5,354,878)

catalyst

methanol

water

rapeseed oil

biodesel

fta   esler   glaero

JO 26.06.96

Figure 5.5 *Example of continuous single feedstock technology.* [38]

they use a solution of sodium methoxide or sodium hydroxide in methanol in order to get very low formation of soaps. After continuous transesterification process, which mostly is conducted in two steps at moderately elevated temperature and ambient pressure the glycerol layers, which are formed at the bottom layer of the reaction mixture, are separated and the raw methyl ester phase is further purified, mainly by different water washing steps. The final product is dried and can be used directly, without distillation, as biodiesel. Most of these biodiesel plants are combined with an oil seed crushing and raffination facility, so it is possible to use high refined oils as starting material. Also these big facilities mainly have an own glycerol purification technology including distillation of raw glycerol into pharmaceutical grade glycerol. The biggest biodiesel production plants mainly are not adapted to use oils with high content of free fatty acids like palm oil or waste oils.

In most single feedstock production plants the glycerol phase is further processed in order to get pharmaceutical grade glycerol. Excess methanol is removed by distillation and the glycerol is distilled under high vacuum and treated with charcoal. The yields for these single feedstock technologies are almost 100 %, because side reactions like saponification are kept to a minimum due to low water and free fatty acid content in the starting material.

5.8.2 Multi Feedstock Technologies

The so-called multi feedstock technologies are capable to process all kinds of various feedstocks, including vegetable oils with higher content of free fatty acids like unrefined oils or palm oil but also waste oils or animal fats. The main difference to single feedstock

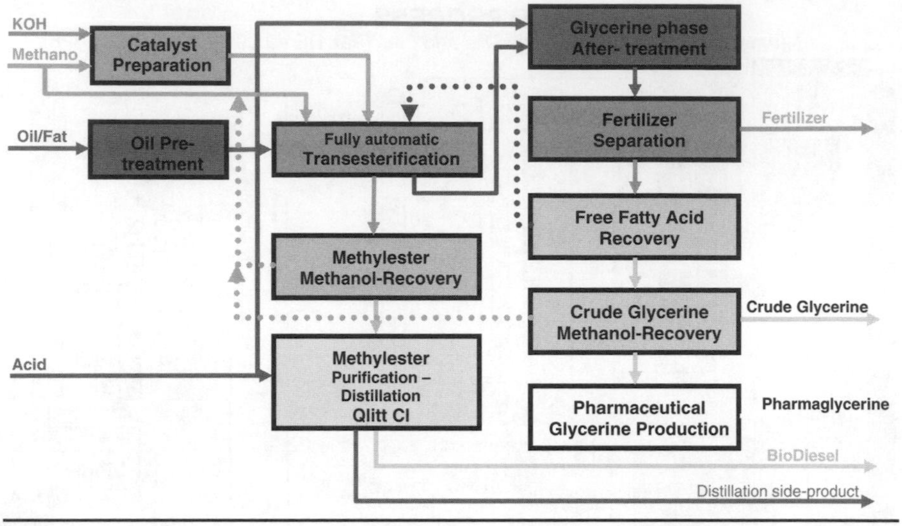

Figure 5.6   *Multifeedstock production scheme according to BioDiesel International.*[39]

technologies is the use of additional reaction steps, like pre-esterification of free fatty acids. So in a first step free fatty acids are pre-esterified with the use of acidic catalysts, followed by one or two alkaline catalyzed transesterification steps. The raw fatty acid methyl esters are purified by water washing steps and additionally can be further refined by vacuum distillation. The main advantage of this technology is the fact that the yield of conversion of fatty acid material into fatty acid methyl esters is almost 100 %. The highest yield can be obtained, when remaining soaps in the glycerol layer are recycled by acidification of the glycerol and separation of free fatty acids, which can be reintroduced into the pre-esterification step or first step of transesterification.[39] Another approach for converting high acidic oils into fatty acid methyl esters is the conversion of fatty acids into glycerides, followed by traditional transesterification.[40]

5.8.3   Small Scale Production Units

A lot of production plants have a production capacity of up to 5000 t/a, using different feedstocks and different production technologies. Mostly these plants have not been built by big biodiesel technology companies, but the technology has been developed by individual groups and organizations based on own experience and development. The glycerol layer must be used directly without any purification, e.g. as substrate for biogas plants, or will be purified to be sold as raw glycerol. The catalyst for transesterifications is mainly potassium hydroxide, because it gives the highest conversion rates. Several of these production plants are organized as co-operatives, using vegetable oils produced locally, and also the biodiesel will be used by the members directly. Most of the very small production units don't have their own facilities for quality control, so the quality of the product might vary and is not guaranteed to meet EN 14214.

5.8.4 Alternative Process Technologies

In order to separate the fatty acid methyl esters during classical transesterification reaction from unreacted emulsified oil, membrane technology has been used producing high-purity fatty acid methyl esters.[40,41,42] Carbon membranes with pore sizes between 0.05 and 1.4 μm were tested and efficient separation of the triglycerides from the FAME reach permeates was achieved. Permeate dephasing occurred at room temperature, the FAME-rich phase contained undetectable levels of glycerol, and the polar, methanol-rich phase can be recycled to the reactor and permit the continuous production of biodiesel. For facilitating the washing steps of biodiesel avoiding the formation of emulsions, hollow fibre membrane extraction was suggested.

References

1. M. Mittelbach, M. Wörgetter, J. Pernkopf, H. Junek (1983) Diesel fuel derived from vegetable oils: preparation and use of rape oil methyl ester, *Energy in Agriculture* **2**: 369–84.
2. *Directive 2003/30/EC of the European Parliament and of the Council of 8 May 2003 on the Promotion of the Use of Biofuels or Other Renewable Fuels for Transport*. http://europa.eu.int/eur-lex/pri/en/oj/dat/2003/l_123/l_12320030517en00420046.pdf
3. Homepage of the European Biodiesel Board: http://www.ebb-eu.org/stats.php
4. Homepage of National Biodiesel Board: http://www.biodiesel.org/pdf_files/fuelfactsheets/Production_Capacity.pdf
5. M. Mittelbach and C. Remschmidt (2006) *Biodiesel, the Comprehensive Handbook*, ed.: M. Mittelbach, Graz. ISBN: 3-200-00249-2.
6. P. Duffy (1852) On the Constitution of Stearine. *Journal of the Chemical Society* **5**: 303–7.
7. G.B. Bradshaw (1941) *Preparation of Detergents*. US Patent 2 360 844.
8. M. Mittelbach and B. Trathnigg (1990) Kinetics of alkaline catalysed methanolysis of sunflower oil. *Fat Science and Technology* **92**: 145–8.
9. B. Freedman, W.F. Kwolek and E.H. Pryde (1986) Quantitation in the analysis of transesterified soybean oil by capillary gas chromatography. *Journal of the American Oil Chemists' Society* **63**(10): 1370–5.
10. M.W. Formo (1954) Ester reactions of fatty materials, *Journal of the American Oil Chemists' Society* **31**(11): 548–59.
11. H. Lepper and L. Friesenhagen (1984) Verfahren zur Herstellung von Fettsäureestern kurzkettiger aliphatischer Alkohole aus freie Fettsäuren enthaltenden Fetten und/oder Ölen.*European Patent EP* 0 127 104 A1.
12. M. Mittelbach, A. Silberholz and M. Koncar (1996) Novel aspects concerning acid catalyzed alcoholysis of triglycerides, *Oils-Fats-Lipids*. Proceedings of the 21st World Congress of the International Society for Fat Research (ISF). The Hague. October 1–6, 1995. Volume 3, 497–9
13. G. Vicente, M. Martínez and J. Aracil (2004) Integrated biodiesel production: a comparison of different homogenous catalysts systems, *Bioresource Technol.* **92**: 297–305.
14. S. Peter and E. Weidner (2007) Methanolysis of triacylglycerols by organic basic catalysts, *Eur. J. Lipid Sci. Technol.* **109**: 11–16.
15. T. Kocsisova, J. Cvengros and J. Lutisan (2005) High-temperature esterification of fatty acids with methanol at ambient pressure, *Eur. J. Lipid Sci. Technol.* **107**: 87–92.

16. J. Graille, P. Lozano, D. Pioch, P. Geneste and A. Guida (1982) Esters méthyliques ou éthyliques comme carburant diesel de substitution, *Oléagineux* **37**(8–9): 421–4.

17. G.J. Suppes, K. Buchwinkel, S. Lucas, J.B. Botts, M.H. Mason and J.A. Heppert (2001) Calcium carbonate catalysed alcoholysis of fats and oils, *Journal of the American Oil Chemists' Society* **78**(2): 139–45.

18. P. Bondioli (2004) The preparation of fatty acid esters by means of catalytic reactions, *Topics in Catalysis* **27**(1–4): 77–82.

19. N. Shibasaki-Kitakawa, H. Honda, H. Kuribayashi, T. Toda, T. Fukumura and T. Yonemoto (2007) Biodiesel production using anionic ion-exchange resin as heterogenous catalyst, *Bioresource Technol.* **98**: 416–21.

20. R. Stern, G. Hillion and J. Rouxel (2000) IFP; US 6,147.196.

21. http://www.ifp.fr/IFP/en/ifp/ab13_02.htm

22. Y.M. Choo and S.H. Ong (1986) Transesterification of fats and oils, *British Patent GB* 2 188 057 A.

23. M. Mittelbach (1990) Lipase catalyzed alcoholysis of sunflower oil, *Journal of the American Oil Chemists' Society* **67**(3): 168–70.

24. W.H. Wu, T.A. Foglia, W.N. Marmer and J.G. Phillips (1999) Optimizing production of ethyl esters of grease using 95 % ethanol by response surface methodology, *Journal of the American Oil Chemists' Society* **76**(4): 517–21.

25. H. Fukuda, A. Kondo and H. Noda (2001) Review. biodiesel fuel production by transesterification of oils, *Journal of Bioscience and Bioengineering* **92**(5): 405–16.

26. R. Uitz (2006) Dissertation, Karl-Franzens-University Graz.

27. M.A. Diasakou, A. Louloudi and N. Papayannakos (1998) Kinetics of the non-catalytic transesterification of soybean oil, *Fuel* **77**(12): 1297–1302.

28. D. Kusdiana and S. Saka (2001) Methyl esterification of free fatty acids of rapeseed oil as treated in supercritic al methanol, *Journal of Chemical Engineering of Japan* **34**(3): 383–7.

29. A. Demirbas (2006) Biodiesel production via non-catalytic SCF method and biodiesel fuel characteristics, *Energy Conversion and Management* **47**: 2271–2.

30. S. Saka, D. Kusdiana and E. Minami (2006) Non-catalytic biodiesel fuel production with supercritical methanol technologies, *Journal of Scientific & Industrial Research* **65**(5): 420–5.

31. D. Kusdiana, S. Saka (2004) Effects of water on biodiesel fuel production by supercritical methanol treatment, *Bioresource Technol.* **91**: 289–95.

32. S. Mahajan, S. Konar and D. Boocock (2006) Standard biodiesel from soybean oil by a single chemical reaction, *J. Amer. Oil Chem. Soc.* **83**: 641–4.

33. D. Fabbri, V. Bevoni, M. Notari and F. Rivetti (2007) Properties of a potential biofuel obtained from soybean oil by transmethylation with dimethyl carbonate, *Fuel* **86**: 690–7.

34. N.E. Leadbeater and L.M. Stencel (2006) Fast, easy preparation of biodiesel using microwave heating, *Energy&Fuels* **20**: 2281–3.

35. K.J. Harrington and C.D' Arcy-Evans (1985) Transesterification in situ of sunflower seed oil. *Ind. Eng. Chem. Prod. Res. Dev.* **24**: 314–18.

36. M.J. Haas, K.M. Scott, W.N. Marmer, T.A. Foglia (2004) In situ alkaline transesterification: an effective method for the production of fatty acid methyl esters from vegetable oils, *J. Amer. Oil Chem. Soc.* **81**: 83–9.

37. N.G. Siatis, A.C. Kimbaris, C.S. Pappas, P.A. Tarantilis and M.G. Polissiou (2006) Improvement of biodiesel production based on the application of ultrasound: monitoring of the procedure by FTIR spectroscopy, *J. Amer. Oil Chem. Soc.* **83**: 53–7.

38. J. Connemann (1993) Verfahren zur kontinuierlichen Herstellung von C1- bis C4-Alkylestern höherer Fettsäuren, *European Patent Application* 0562504A2.

39. M. Mittelbach and M. Koncar (1994) Process for preparing fatty acid alkyl esters. *European Patent EP* 0 708 813 B1.
40. F. Luxem, J. Galante, W. Troy and R. Bernhardt (2006) *US* 7.087.771.
41. P. Cao, A. Tremblay, M. Dubé and K. Morse (2007) Effect of membrane pore size on the performance of a membrane reactor for biodiesel production, *Ind. Eng. Chem. Res.* **46**: 52–8.
42. H.Y. He, X. Gu O and S.L. Zhu (2006) Comparison of membrane extraction with traditional extraction methods for biodiesel production, *J. Amer. Oil Chem. Soc.* **83**: 457–60.

6

Bio-based Fischer-Tropsch Diesel Production Technologies

Robin Zwart

Energy Research Centre of the Netherlands Biomass, Coal and Environmental Research Petten, The Netherlands

René van Ree

Wageningen University and Research Centre, Wageningen, The Netherlands

6.1 Introduction

Bio-based transportation fuels are expected to contribute significantly to the future transportation fuel portfolio both on national, EC and global levels. Bio-based transportation fuels that are currently produced and used are so called *first generation biofuels*. Examples are: pure vegetable plant oils, biodiesel produced from the seeds of oil-rich crops and from waste vegetable oils and animal fats, conventional bioethanol/ETBE produced from sugar and starch crops, and upgraded biogas produced from the digestion of organic residues. An advantage of these first generation biofuels is that production technologies are commercially available and that these fuels are already being produced for, and applied mainly as blending agents in, the current transportation fuel market. Disadvantages are (1) that the overall CO_2-reduction potential of these fuels compared to their fossil alternatives, taking into account the whole biomass–fuel application chain, is generally reported to be less than 50 %, and (2) that the production processes are relatively raw material specific, decreasing the overall market application potential.

Currently, technologies are being developed for the production of so called *one-and-a-half generation biofuels*, which have better properties. Examples are: the upgrading (hydrogenation) of biodiesel to a higher-quality bio-based diesel, the production of higher alcohols (i.e. biobutanol) from sugar and starch crops, and the production of bioethanol or biobutanol from a wider range more difficult to convert raw materials.

Considering the European Policy goals on the implementation of biofuels, i.e. 2 % and 5.75 % fossil fuel substitution on energy basis in 2005 and 2010, these have to be met fully by the implementation of additional first and one-and-a-half generation biofuel production capacity. To meet to the longer-term market demands, for example the 25 % fossil fuel substitution directive for 2030, as mentioned in the Vision document of the European Technology Platform on Biofuels, and to gradually shift to biofuels with a better overall CO_2-reduction potential, the introduction of so called *second generation biofuels* is a necessity. Examples are: bioethanol or biobutanol produced from lignocellulosic-rich raw materials and Biomass-to-Liquids (BtL) products, like Fischer-Tropsch (FT) diesel, dimethylether (DME), bioSNG, bioCNG and biomethanol.

This chapter will fully concentrate on the current technological status and market perspectives for the production of FT-diesel from lignocellulosic-rich raw materials (wood, straw, etc.). Aspects that will be described in more detail are: the theoretical background of catalytic FT-diesel synthesis and the techno-economic aspects of biomass-based integrated gasification-based FT-diesel production concepts considered in some major demonstration projects.

6.2   Theoretical Background Catalytic FT-Diesel Synthesis Process

The Fischer-Tropsch (FT) synthesis was discovered in 1923 by the German scientists F. Fischer and H. Tropsch at the Kaiser Wilhelm Institute for Coal Research in Mülheim, Germany. In the synthesis hydrocarbons are produced from syngas, *viz.* a mixture of the gases CO and H_2. Historically, FT-processes have been operated on a large industrial scale to produce synthetic fuels as alternative for non-available fossil fuels (i.e. in Germany in the 1930s and 1940s, and in South Africa during the oil boycott). To date, the FT-process receives much attention because the hydrocarbon products are 'ultra-clean' due to the nature of the synthesis process, i.e. they are essentially free of sulphur and aromatics. Shell Gas-to-Liquids (GtL) derived FT-diesel blended with fossil diesel (i.e., V-power) is available to reduce local soot and SO_2 emissions.

6.2.1   Chemistry

6.2.1.1   Synthesis

In the catalytic FT-synthesis one mole of CO reacts with two moles of H_2 to form mainly paraffin straight-chain hydrocarbons (C_xH_{2x}), with minor amounts of branched and unsaturated hydrocarbons (*i.e.* 2-methyl paraffins and a-olefins), and primary alcohols. Typical operation conditions for FT-synthesis are temperatures of 200–250 °C and pressures between 25 and 60 bar.[1] In the exothermic FT-reaction about 20 % of the chemical energy is released as heat:

$$CO + 2H_2 \Rightarrow -(CH_2)- + H_2O$$

6.2.1.2 Catalysts

Several types of catalysts can be used for the FT-synthesis; the most important are based on iron (Fe) or cobalt (Co). Cobalt catalysts have the advantage of a higher conversion rate and a longer life (over five years). The Co catalysts are in general more reactive for hydrogenation and produce therefore less unsaturated hydrocarbons (olefins) and alcohols compared to Fe catalysts. Iron catalysts have a higher tolerance for sulphur, are cheaper, and produce more olefin products and alcohols. The lifetime of the Fe catalysts is short and in commercial installations generally limited to eight weeks.

The FT-reaction consumes hydrogen and carbon monoxide in a molar ratio of $H_2/CO = 2$. When the ratio in the feed gas is lower, it can be adjusted with the water-gas shift (WGS) reaction:

$$CO + H_2O \Leftrightarrow CO_2 + H_2$$

Iron-based FT-catalysts show considerable WGS activity and the H_2/CO ratio is adjusted in the synthesis reactor. In the case of cobalt-based catalysts the ratio needs to be adjusted prior to FT-synthesis.

6.2.1.3 Product Distribution

The polymerization-like FT chain-growth reaction results in a range of products, comprising light hydrocarbons (C_1-C_2), LPG (C_3-C_4), naphtha (C_5-C_{11}), diesel (C_{12}-C_{20}), and wax ($>C_{20}$) fractions. The theoretical chain length distribution can be described by means of the Anderson-Schulz-Flory (ASF) equation, which is represented as:

$$\log\frac{W_n}{n} = n \cdot \log\alpha + \log\frac{(1-\alpha)^2}{\alpha}$$

where W_n is the weight fraction of a product consisting of n carbon atoms, and a the chain growth probability factor. Higher α values give more high-molecular weight products as can be seen in Figure 6.1. The value of α is characteristic of the particular catalyst employed in the FT-process and, depending on the needs of a particular production process, catalysts and process operation conditions can be tailored towards the production of predominantly low or higher molecular weight hydrocarbons.

In practice, there is often a deviation from the ideal ASF distribution especially with regards to the lower hydrocarbon yields. C_1 yields are usually higher than predicted, whereas C_2 (as well as C_3 and C_4) yields are lower. To incorporate the deviation from the ideal ASF distribution with regard to the yields of the C_1-C_4 hydrocarbon, as 'rule-of-thumb' these values can be recalculated according to:

$$W_1 = \frac{1}{2} \cdot \left(1 - \sum_{i=5}^{\infty} W_i\right) \qquad W_{2,3,4} = \frac{1}{6} \cdot \left(1 - \sum_{i=5}^{\infty} W_i\right)$$

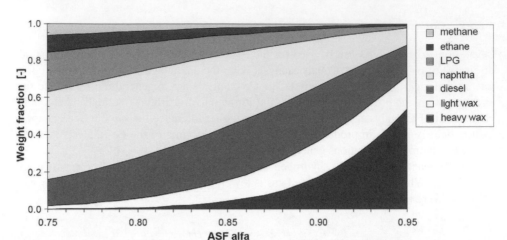

Figure 6.1 *Theoretical ASF product distribution for FT-synthesis, with LPG (C_3-C_4), naphtha (C_5-C_{11}), diesel (C_{12}-C_{20}), light wax (C_{21}-C_{30}), and heavy wax ($>C_{30}$). Reproduced by permission of ECN.*

6.2.2   Feed Gas Specifications

The catalysts used in FT-synthesis are intrinsically very sensitive to small amounts of impurities. In commercial operation, catalysts are replaced or regenerated after a certain operational period. The definition of the gas cleaning is therefore based on economic considerations: investment in gas cleaning versus accepting decreasing production due to poisoning of the catalyst. Therefore, there are no 'hard' data on maximum levels for impurities in FT feed gas. For each plant the acceptable levels may be different. Rule-of-thumb specifications are presented in Table 6.1 for known impurities, and for impurities that might be present in biomass-derived gases.[2,3]

A maximum value of less than 1 ppmV is defined for both the sum of the N_2-containing and S-containing compounds. For the halides and alkaline metals a lower level of less than 10 ppbV is assumed. With respect to the organic constituents that are present in biomass product gases (i.e. tars and BTX), tars in general, there are no limits regarding poisoning

Table 6.1 *FT feed gas specifications. Reproduced by permission of ECN*

Impurity	Removal level
H_2S + COS + CS_2	< 1 ppmV
NH_3 + HCN	< 1 ppmV
HCl + HBr + HF	< 10 ppbV
alkaline metals	< 10 ppbV
Solids (soot, dust, ash)	essentially completely
Organic compounds[α] (tars)	below dew point
- *class 2*[β] *(hetero atoms)*	< 1 ppmV

[α] organic compounds include also benzene, toluene and xylene (BTX).
[β] class 2 tars comprise phenol, pyridine, and thiophene.

of the catalyst. However, as the gas needs to be compressed to 25–60 bar for FT-synthesis, the concentration of the organic compounds must be below the dew point at FT pressure to prevent condensation and fouling in the system. Specifically, class 2 tars with S or N hetero atoms (e.g. thiophene and pyridine) need to be removed below ppmV level, as they are intrinsically poisonous for the catalyst. Solids must be removed essentially completely, as they foul the system and may obstruct fixed-bed reactors.

With respect to the other possible constituents (depending on the gasification concept) of the FT feed, i.e. CO_2, N_2, CH_4, and larger hydrocarbons, there are no hard specifications. However, similar to the gas cleaning, specifications are set by economic considerations. For the concentration of these gases, which are inert in the FT-synthesis, a soft maximum of 15 vol % is defined (but the lower, the better). The presence of inert components requires larger reactors and higher total gas pressures. CO_2 can readily be removed with standard techniques, but N_2 and the light-end hydrocarbons cannot be removed at reasonable costs. Therefore, in the production of the FT feed gas the presence of significant concentrations of the latter compounds should be avoided.

6.2.3 Commercial Processes

6.2.3.1 Fischer-Tropsch Synthesis

Today, FT-synthesis is an established technology[4-6] and two companies have already commercialized their FT-technology, i.e. Shell (1st plant in Malaysia) and Sasol (several plants in South Africa), using natural gas and coal as feedstock to produce the syngas, respectively. Sasol uses iron catalysts and operates several types of reactors, of which the slurry bubble column reactor is the most versatile (i.e. applied in the Sasol Slurry Phase Distillate (SSPD)). Shell operates the Shell Middle Distillate Synthesis (SMDS) process in Bintulu, Malaysia, which produces heavy waxes with a cobalt-based catalyst in multi-tubular fixed bed reactors. Shell has started the construction for a 140,000 bbpd SMDS plant in Qatar, while Sasol has started a 34,000 bbpd cobalt-based SSPD plant in 2007, also in Qatar.

6.2.3.2 Syngas Production and Clean-up

Most syngas is produced by partial oxidation of natural gas (84 %); syngas can also be produced by gasification of coal, while small amounts are generated in refinery processes. The cleaning of the raw syngas from partial oxidation is a well-known and commercially available process.[7] The general approach is to quench the raw hot gas with water to cool the gas and removed solid particles (viz. dust, soot, and ash) and the volatile alkaline metals. Upon syngas production, H_2S, NH_3, COS and CS_2, and HCN are formed from sulphur and nitrogen in the fuel. The NH_3 is removed downstream together with the halides (*viz*. HCl, HBr, and HF) with a water scrubber and H_2S is removed either by absorption or after conversion to elementary sulphur (i.e. the Claus process). The adsorption removal is preferred when relatively small amounts of H_2S are present. Similar is valid for the presence of COS and HCN. These impurities are hardly removed in the gas cleaning and are captured in the guard beds. When a syngas contains higher loads of these compounds it is economically more attractive to install a hydrolysis step to convert them to H_2S and NH_3, respectively. These components are readily removed in the gas cleaning. With this cleaning process the syngas specifications are met (Table 6.1).

6.3   Biomass Gasification-Based FT-Diesel Production Concepts

6.3.1   Syngas and FT-diesel Market

Syngas is a versatile building block in chemical industry.[8,9] The total global annual use of fossil-derived syngas is approximately 6000 PJ_{th}, which corresponds to 2 % of the total primary energy consumption. The largest part of the syngas is used for the synthesis of ammonia for fertilizer production ($\sim$55 %), the second largest share is the amount of hydrogen from syngas consumed in oil refining processes ($\sim$24 %), and smaller amounts are used for methanol production (12 %). Figure 6.2 shows the present and predicted future syngas market distribution.[10] Today's, global use of syngas for the production of transportation fuels in the so-called 'gas-to-liquids' processes (GTL) correspond to approx. 500 PJ per year, i.e. from the FT-processes of Sasol in South Africa and Qatar and of Shell in Bintulu, Malaysia.

6.3.1.1   Scale and Location

The future biosyngas demand exceeds the present syngas consumption by a factor of eight. Therefore, it is clear that large biosyngas production capacities are needed to meet the European and national renewable energy and CO_2-emission reduction targets. Not only are large installed capacities necessary, also the individual plants, compared to typical biomass plants, have to be large considering the typical plant scales for the two main applications, i.e.

- Transportation fuels in BTL plants: few 100 MW to several 1,000 MW
- Chemical sector: 50 $\sim$ 200 MW

Syngas demands for liquid fuel synthesis will typically be larger than 1000 MW for plants, where the whole chain from biomass to the final product is realized, to benefit from

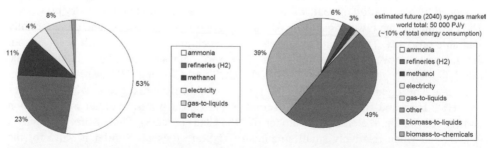

Figure 6.2   *Present world syngas market (left) and predicted world syngas market in 2040 (right). Reproduced by permission of ECN.*

economy of scale, which is necessary to reduce costs. For illustration, the Shell GTL plant in Malaysia of 12,500 bbpd (*i.e.* ~1000 MW) is considered as a demonstration plant, while the new plant in Qatar will have a six times higher capacity (75,000 bbpd or ~6000 MW). Another possibility is that there are several smaller plants in the size of ~500 MW, which produce only intermediate products, e.g. raw liquid products, where the final work-up is done in a central facility.

The typical syngas demands for chemical processes correspond to 50–200 MW_{th}. Even though the scale of an individual biosyngas plant may be relatively small, in most cases the plant will be part of a larger centralized chemical infrastructure with several other processes and plants to optimize energy and product integration (i.e. the syngas consumer). There is only a limited market for stand-alone small-scale biosyngas production for distributed chemical plants; although there will always be exceptions.

To ensure cost-effective biomass supply (i.e. avoid land transport) biosyngas production plants will be constructed close to ports or larger waterways. For the selection of the location the same considerations apply as for current coal-fired power plants and their coal logistics. Also the main large concentrations of chemical industry are located on locations easy accessible from water, e.g. the Dutch Maasvlakte near Rotterdam, and the German Ruhrgebiet.

6.3.1.2 Implementation

As a large total installed biosyngas production capacity with large individual plants is required to meet the ambitious renewable energy targets,[11] a robust, fuel-flexible, and high-efficient technology for optimum biomass utilization is required to guarantee availability. In developing BtL technology, two possible routes can be followed, see Figure 6.3.

The first route comprises up-scaling of the small and medium scale biomass-based gasification technologies that are currently mostly used for distributed heat and power (CHP) production. In this route it will take a long time before a significant biosyngas production capacity is installed. Either, a large number of plants will have to be put in operation or the technology has to be up-scaled, which will take an additional development

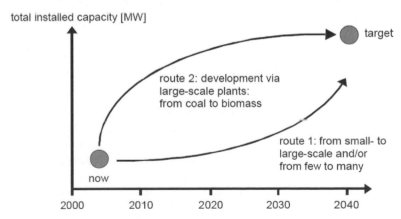

Figure 6.3 *Roadmap to reach for large-scale implementation of biosyngas and BtL products. Reproduced by permission of ECN.*

period of a decade. Therefore, it is questionable if the ambitious renewable energy targets can be met by following this route.

The second and perhaps to be preferred route comprises adapting today's large-scale coal-based (entrained flow (EF)) gasification technology, initially by co-firing biomass and later on construct biomass-based gasifiers. In this way the installed biosyngas capacity can be increased rapidly, as basic technology is already proven on large scale.

6.3.2 Biomass Feedstock

Approximately 50 % of the biomass globally available for energy purposes (i.e. the technical potential) is wood or wood residues. A further 20 % is strawlike, which share will increase to 40 % when strawlike crops are selected as energy crops.[12] Hence, for utilization of the large amounts of biomass for biosyngas production, it is important to develop technical routes that are able to utilize both wood and straw materials. Biomass materials like manure and waste streams will play only a minor role in the biosyngas production, as the absolute amounts of these streams available for biosyngas production are very low compared to the required total amounts of biomass. Therefore, they are not of significance for large-scale biosyngas production.

Due to the distributed and global generation of the biomass, large transport distances are unavoidable. Transport by truck is the major cost driver in biomass transport; therefore, the transport over land should be minimized.[2] Strawlike materials have a much lower bulk energy density, which would result in higher transport and transhipment costs. Therefore, energy densification of straw is desired to reduce transport costs and allow easier handling, *viz.* grasses and straw are converted into a bioslurry via flash pyrolysis,[13] as will be discussed in Sections 6.3.4.4 and 6.3.4.5. Wood, although already relatively high-energy dense, will preferable also be densified (e.g. by torrefaction discussed in Section 6.3.4.2 or again pyrolysis, resulting also in simplification of the biomass feeding system of gasification) before being transported by ship to large centralized conversion facilities.

The transition to green alternatives requires biomass, which should be available in large quantities. Since wood and grasslike material make up 70–90 % of the total technically available amount of biomass world-wide, it is reasonable to focus on these biomass fuels as main renewable energy sources for chemicals and fuels.[14]

6.3.3 Syngas Production Technology

In order to produce biosyngas cost-effectively, high biomass-to-syngas yields are required. This implies that upon gasification of biomass the maximum share of energy contained in the biomass should be converted into the syngas components H_2 and CO.

There are two thermo-chemical ways to produce synthesis gas (H_2 and CO) from biomass: either by applying high temperatures or by using a catalyst at a much lower temperature Figure 6.4.[15] The first route includes a fluidized bed gasifier and a downstream catalytic reformer, both operating at approximately 900 °C. In product gas from low temperature gasification the syngas components H_2 and CO typically contain only ~50 % of the energy in the gas, while the remainder is contained in CH_4 and higher (aromatic)

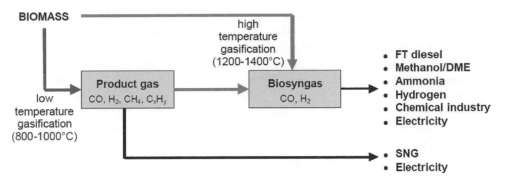

Figure 6.4 *Two biomass-derived gases via gasification at different temperature levels. Reproduced by permission of ECN.*

hydrocarbons that have to be catalytically reformed. The second route generally requires temperatures as high as 1300 °C and generally involves an entrained flow gasifier. Upon high temperature gasification (>1300 °C) all the biomass is completely converted into biosyngas. Biosyngas is chemically similar to syngas derived from fossil sources and can replace its fossil equivalent in all applications. The two options concern two different evolution trajectories. The following paragraphs cover the two options.

6.3.3.1 Fluidized Bed Gasification with Catalytic Reformer

Fluidized bed gasification of biomass presently is a common way of converting biomass. Many different technologies are available. The air-blown circulating fluidized bed (CFB) is the most common one. Most fluidized bed applications involve close-coupled combustion with little or no intermediate gas cleaning. Power and/or heat are the usual end products. The gas produced by a fluidized bed gasifier (typically operated at 900 °C) contains H_2, CO, CO_2, H_2O, and considerable amounts of hydrocarbons like CH_4, C_2H_4, benzene and tars. Although this so-called product gas is suitable for combustion processes, it does not meet the requirements of synthesis gas, which is needed to produce biofuels or chemicals. The product gas needs further treatment in a catalytic reformer where hydrocarbons are converted into H_2 and CO (and CO_2 and H_2O). Since most syngas to liquid fuels conversion processes require a raw gas with very little or no inert gases, gasification and reforming should apply pure oxygen instead of air. Steam is usually added as a moderator. Another option to avoid N_2 dilution is to use an allothermal or indirect gasifier. In these reactors, gas production and heat generation do not take place in the same reactor. This enables the use of air (in the heat generating reactor), without having the N_2-dilution of the gas coming from the gas generation reactor. Examples of indirect gasifiers are: the SilvaGas-process developed by Battelle in the US,[16] the MTCI-process,[17] the MILENA developed by the Energy research Centre of the Netherlands (ECN),[18,19] and the FICFB-concept developed by the Vienna University[20] and in operation at the Biomass-CHP in Güssing. The FICFB based biomass-CHP plant in Güssing is one of the facilities (as well as the Choren and Chemrec plants discussed in Sections 6.3.4.5 and 6.3.4.6) have been used in the EU-funded project RENEW. Within this project production routes for BTL fuels have been demonstrated and the full supply chain was assessed in terms of biomass potential, life-cycle, costs and technological options. Fuels were be produced and tested in order to demonstrate benefits of optimized fuels for advanced power trains.

In the EU-funded project Chrisgas, the existing 18 MW_{th} pressurized CFB-gasifier in Värnamo will be refurbished to produce syngas.[21] This includes operation on oxygen/steam instead of air, the installation of a high temperature filter, a catalytic reformer, and a shift reactor. In five years time, the plant should produce 3500 m_n^3/h H_2 and CO at 10 bar. The project is carried out by the VVBGC consortium (Växjö Värnamo Biomass Gasification Centre). In the next phase, fuel synthesis will be added to the plant.

Another initiative in this category is by VTT that advocates fluidized bed gasification as the process to generate clean fuel gas as well as syngas from biomass. Fuel flexibility is considered the major advantage. VTT started the UCG-programme (Ultra Clean Fuel Gas) and a 500 kW_{th} PDU is operating.[22] It consists of a pressurized fluidized bed, catalytic reforming, and further cleaning and conditioning. The catalytic reformer is meant to reduce hydrocarbons (benzene and larger) completely and methane by over 95 %. The test unit will support RD&D focusing on methanol and FT-diesel production via syngas as well as the production of SNG, H_2, and electricity by fuel cells. The present estimate of a 300 MW_{th} plant based on above described VTT process show that FT-diesel and methanol can be produced for approximately 12 €/GJ (feedstock price 2.8 €/GJ). A plant this size can be largely constructed as a single train.

The German institute CUTEC has constructed an oxygen-blown 0.4 MW_{th} CFB gasifier connected to a catalytic reformer. Part of the gas is compressed and directed to a FT-synthesis reactor.[23] Apart from these above-mentioned initiatives to develop technology to produce syngas by fluidized bed gasification and catalytic reforming, many others apply catalytic reforming reactors for gas conditioning. This, however, generally focuses on the catalytic reduction of large hydrocarbon molecules (*viz.* tars). Reforming of methane usually is not one of the goals in these concepts.

6.3.3.2   *Entrained Flow Gasification*

The non-catalytic production of syngas (H_2 and CO) from biomass generally requires high temperatures, typically 1300 °C. The most common reactor for this is the entrained flow gasifier.[24] Since biomass contains mineral matter (ash), a slagging entrained flow gasifier seems to be the most appropriate technology.[25] Entrained flow reactors need very small fuel particles to have sufficient conversion. This requires extensive milling of solid fuels, which is energy intensive and generally produces particles that cannot be fed by conventional pneumatic systems.[25] R&D therefore focuses on ways to technically enable the fuel feeding, as discussed in Section 6.3.4, as well as on the improvement of the economics of the whole chain. The most promising pre-treatment options are torrefaction and pyrolysis. These options enable efficient and cheap production of syngas from biomass, mainly because it is characterized by relatively cheap (long-distance) transport.

Different slagging entrained flow gasifiers are operated worldwide, but only few have experience with biomass. Former Future Energy, now Siemens, in Freiberg Germany commercializes entrained flow gasifier technology for biomass, waste, and other fuels.[26] It owns a 3 MW_{th} pilot plant that has been operated with many different biomass fuels.

Furthermore, Siemens supplied the 120 MW_{th} entrained flow gasifier, which is commercially operated on waste material in the Schwarze Pumpe complex in Germany.[27] Another example is the Buggenum IGCC-plant in the Netherlands where biomass is co-gasified with coal in a slagging entrained flow gasifier. Tests have been conducted using up to 34 wt % biomass in the feeding mixture.[28] This biomass was mainly sewage sludge and chicken manure, not generating pulverizing problems and feeding problems when mixed with coal. Also a mix of coal and wood dust has been tested successfully.

6.3.3.3 Polygeneration

The fluidized bed gasification with catalytic reforming and entrained flow gasification processes, described above, focus on the production of syngas with high yields of H_2 and CO. This is desirable in order to get the maximum production efficiency of biofuels/chemicals like methanol and FT-diesel. The alternative approach is called polygeneration. In this case, the H_2 and CO from a gas are used for the (once-through) synthesis of a biofuel/chemical, and the remaining components in the gas are used in a different way, e.g. the production of power. The waste gasification plant of the Schwarze Pumpe complex in Germany is an example where waste is converted into a product gas containing considerable amounts of hydrocarbons. The gas is cleaned and used as feedstock to produce methanol. The remaining gas (mainly methane) is used as fuel for a 75 MW_e combined-cycle to produce power.[27, 29]

6.3.3.4 Biorefinery

Another integrated biofuel concept is sometimes referred to as biorefinery, a concept that is of great interest in the US where conventional (biological) fermentation is combined with thermo-chemical conversion with syngas as intermediate product. This concept efficiently produces ethanol and other alcohols from different kinds of biomass. This so-called advanced ethanol refinery plant is expected to produce alcohols for less than 1 \$/gallon.[30]

6.3.3.5 The Optimum Syngas Production Technology

The pressurized oxygen-blown entrained flow (EF) gasifier is considered to be the heart of the optimum large-scale syngas/FT biodiesel production process, as presented in Figure 6.5. This technology was identified as optimum technology for biosyngas production as it has the advantages of (i) high efficiency to biosyngas, (ii) fuel flexibility for all types of biomass, e.g. wood, straw, and grassy materials, (iii) suitability for scales of several hundreds to a few thousand megawatt, and (iv) possibility to operate on coal as back-up fuel.

As the EF gasification for coal is a well-established and commercial available technology, and has been demonstrated to be able to co-fire significant amounts of biomass,[28] it also enables a short-term graduate transfer from coal-to-liquid (CTL) plants to biomass-to-liquid (BTL) plants.

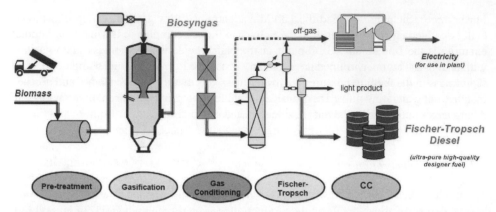

Figure 6.5 *Schematic line-up of the optimum syngas/FT biodiesel production process. Reproduced by permission of ECN.*

6.3.4   Biomass Feeding and Pre-treatment

Biomass is different from coal in many respects; the most relevant relates to feeding. Biomass requires significant pre-treatment to allow stable feeding into the gasifier without excessive inert gas consumption. In addition to the requirement to pre-treat the biomass for feeding, it may also be desired for purpose of densification of the material. Due to the smaller volume transport costs are reduced and the stability of the gasifier operation is increased, due to the higher energy density of the feed.

The development of new approaches to biomass feeding and pre-treatment is necessary to reach high biomass-to-biosyngas and overall system efficiencies in biomass-fired entrained flow gasifier systems.

In Figure 6.6 four possible specific pre-treatment and feeding options are shown for different biomass streams. A fifth feeding option, not presented in this figure, is feeding a residual product, e.g. black liquor from pulp and paper industry, and hence integrate the

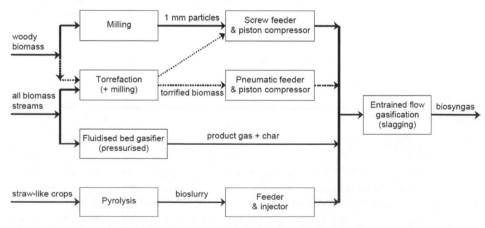

Figure 6.6 *Different biomass pre-treatment and feeding option. Reproduced by permission of ECN.*

syngas production in existing industries. The pre-treatment and feeding options are not competitive but complementary or an alternative to each other.

6.3.4.1 Milling and Screw Feeding

The biomass-to-biosyngas and the overall system efficiency is optimized when no pre-treatment is necessary, i.e. as milling wood down to 1 mm wood particles is sufficient to reach complete conversion. Then the electricity consumption is not excessive and no separate pre-treatment process is required with the accompanying loss in efficiency and the investment and running costs. (This option is not applicable to straw, as knots in the straw cuttings are not completely converted.)

A piston compressor is used for pressurization. Development of a new feeding and dosing system is a necessity, as pneumatic feeding is not possible due to the plugging nature of the fibrous biomass. This route affords higher biomass-to-biosyngas energy and system efficiencies because of the low electricity and inert gas consumption. Therefore, this route is the preferred option, especially for expensive biomass. However, conditionally that (i) the 1 mm biomass cuttings can be fed by a screw to the gasifier and (ii) that sufficient conversion is achieved in the gasifier.[25]

6.3.4.2 Torrefaction and Pneumatic Feeding

Torrefaction is a mild heat treatment at 250–300 °C that efficiently turns solid biomass into a brittle, easy to pulverize material ('bio-coal') that can be treated as coal.[32,33] Furthermore, torrefied biomass can be pelletized very easily to obtain a dense and easy to transport biomass fuel.[34] The hydrophobic nature of torrefied material further simplifies logistics. Pulverized torrefied biomass can be fed like coal, thus enabling a smooth transition from coal to biomass.

Although torrefaction is a rather common process, in e.g. the coffee industry, it has never been optimized for efficient production of a brittle 'bio-coal'. Research at the Energy Research Centre of the Netherlands (ECN) shows, that the conversion of wood into a torrefied wood with similar milling characteristics as hard coal can have an energetic efficiency of 90–95 % LHV. The gases produced during torrefaction can be used to supply the thermal needs of the process.

6.3.4.3 Fluidized Bed Gasification

High overall biomass-to-biosyngas and system efficiencies can also be obtained with a system in which a (pressurized) fluidized bed gasifier is used to 'pretreat' the biomass (i.e. 78 and 85 %). The raw product gas, containing hydrocarbons and tars as well as the unconverted char and some bed material from the bed, is directly fed into the entrained flow gasifier. Upon entrained flow gasification all the organic compounds and char are converted into syngas, therefore, the product gas quality and the carbon conversion of the fluidized bed gasifier are irrelevant. The entrainment of bed material from the fluidized bed gasifier is no problem and even preferred, as it will act as flux for the entrained flow gasifier. Major advantage is that no pre-treatment of the biomass is necessary (chips of

5 cm are acceptable) and that all types of biomass can be processed (i.e. both woody and straw-like biomass). However, this system is the most challenging, as the entrained flow gasifier requires a very stable feed flow to guarantee safe operation, while a fluidized bed gasifier typically has some variations in the product gas flow.[25]

6.3.4.4   Fast Pyrolysis for Bioslurry Production and Feeding

Pyrolysis takes place at approximately 500 °C and can convert solid biomass into a liquid product (bio-oil) in a process that is called flash-pyrolysis. The conversion efficiency will increase to 90 % by including char in the oil to produce a bio-slurry.[35,36] Slurries can be pressurized and fed relatively easily.

The Forschungs Zentrum Karlsruhe (FZK) developed a concept to produce syngas from agricultural waste streams like straw.[36] In this concept, straw is liquefied locally by flash pyrolysis into oil and char slurry, which is subsequently transported and added to a large pressurized oxygen-blown entrained flow gasifier. This approach offers the advantage of low transport costs of the energy dense slurry and large-scale syngas production and synthesis. At the same time, the problem of pressurizing biomass is solved, since slurries are pumpable. An important patented feature of the concept is formed by the fact that milling of char can turn a solid mass into liquid slurry by eliminating the volume of the pores of the char.[37] Flash pyrolysis plants typically will be 100 MW$_{th}$ input capacity. FZK developed the Lurgi-Ruhrgas concept that includes twin screws for pyrolysis. A 5–10 kg/h PDU as well as a 500 kg/h pilot plant are available at the premises of FZK. Several slurries have been tested in the 500 kg/h entrained flow gasifier of Siemens in Freiberg to study its gasification and slagging behaviour.[38] Slurries from straw have been successfully converted into syngas with high conversion and near zero methane content. Biofuel costs produced by this kind of biomass-to-liquid plant will be around 1 €/kg (approximately 23 €/GJ), based on a feedstock price of 3 €/GJ straw.[36]

6.3.4.5   Slow Pyrolysis for Char and Gas Production

Choren develops the Carbo-V concept where solid biomass is pre-treated by slow pyrolysis to yield char and gases instead of the fast pyrolysis slurry.[39] The gases are gasified at high temperature (typically 1300 °C) to generate syngas. The char is pulverized and injected downstream the high-temperature reactor in order to cool the syngas by endothermic char gasification reactions. This so-called chemical quench cools the syngas to approximately 1000 °C. The concept has been demonstrated in the 1 MW$_{th}$ alpha plant in Freiberg, Germany. Since 2003, biofuel synthesis has been added to the plant. After a short period of methanol synthesis, the unit was modified to FT-synthesis. Late 2002, Choren started the construction of the 45 MW$_{th}$ 5 bar beta plant. End 2008, this plant is projected to produce FT-diesel with approximately 50 % thermal efficiency from wood. The next plant is planned for Schwedt, converting 1 million ton per year.

6.3.4.6   Black Liquor Utilization

Existing pulp and paper industry offers unique opportunities for production of biofuels with syngas as intermediate. An important part of many pulp and paper plants is formed by the chemicals recovery cycle where black liquor is combusted in so-called Tomlinson boilers. Substituting the boiler by a gasification plant with additional bio-fuel and electricity production is very attractive, especially when the old boiler has to be replaced. The economic calculations are based on incremental costs rather than absolute costs. This method seems generally acceptable[40–43] and leads to e.g. methanol production costs of 0.3–0.4 €/litre of petrol equivalent.[41–43] It must be realized that biofuel plants, which are integrated in existing pulp and paper mills should match the scale of the paper mill. The integrated capacity of the biofuel plant therefore is typically 300 MW_{th},[41] which is at least 10 times smaller than commercial fossil fuel based methanol and FT-plants. Efficiencies of biofuel plants, which are integrated in pulp- and paper industry, are often reported based on additional biomass, which is needed to produce the additional biofuels. This results in very high reported values of 65 % to even 75 %.[42,43]

Chemrec develops a technology needed to convert black liquor into initially syngas and subsequently produce biofuels like DME, methanol, etc.[42] It is a dedicated entrained flow gasifier operated at temperatures as low as 1000–1100 °C. This is possible due to the presence of large amount of sodium, which acts as a gas-phase catalyst in the gasifier. Chemrec constructed DP1, a 3 MW_{th} entrained flow gasifier operating at 30 bar in Piteå, Sweden, next to the Kappa paper mill. It includes gas cooling by water quench and gas cooler. The syngas will have a composition (vol % dry) of approximately: 39 % H_2, 38 % CO, 19 % CO_2, 1.3 % CH_4, 1.9 % H_2S, and 0.2 % N_2. At present, the pilot plant is full-time operated. Demonstration plant DP2 will be a black liquor gasification combined cycle (BLGCC) plant in Piteå. DP3 will be located in Mörrum. This will produce a biofuel. Both plants will be constructed in 2006/2007. Implementation of such concept in the US paper and pulp industry could produce 4.4 % of current US petroleum/diesel consumption.[40] In Finland and Sweden this could be significant, as high as 51 % and 29 % of the respective national use of transportation fuels.[43]

6.3.5 Syngas Cleaning and Conditioning

Generally, gasification technologies are selected for their high efficiency to produce H_2 and CO, with little or no hydrocarbons. The presence of minor impurities (soot, sulphur, chlorine, and ammonia) will however be inevitable. Since the concentration of these components generally exceed the specification of a catalytic synthesis reactor (as presented in Table 6.1), gas cleaning is necessary. It must be realized that there is an economic trade off between gas cleaning and catalyst performance. Cleaning well below specifications might be economically attractive for synthesis processes that use sensitive and expensive catalytic materials. Since raw bio-syngas resembles syngas produced from more conventional fuels like coal and oil residues, gas cleaning technologies will be very similar. This means that it most probably will include a filter, Rectisol unit, and downstream gas polishing to remove the traces. This involves e.g. ZnO and active carbon filtering.

Because the H_2/CO-ratio generally needs adjustment, a water-gas-shift reactor will also be part of gas conditioning. The Rectisol unit then combines the removal of the bulk of the impurities and the separation of CO_2.

6.4 Economics of Biomass-Based FT-Diesel Production Concepts

In the previous paragraph pressurized oxygen-blown entrained flow gasification is considered to be the heart of the optimum large-scale syngas/FT-diesel production process. In this paragraph, the production costs of the BTL diesel fuel are discussed. The fuel production costs are composed of the costs for the biomass feedstock material, transport, transshipment, and storage, pre-treatment, and the conversion (gasification, cleaning, synthesis, and product upgrading).[44] The schematic line-up of the integrated biomass gasification and Fischer-Tropsch synthesis (BTL) plant is shown in Figure 6.5.

It is assumed that the BTL plant is located in the centre of a circular forest area. Of the area 38 % is production forest of which 50 % is exploitable with an annual biomass production yield of 10 ton dry solids per hectare. The radius of the area depends on the scale of the BTL plant, i.e. on the amount of biomass feedstock required. The biomass is assumed to be chipped and dried (7 % moisture; bulk density 202 kg/m^3; calorific value LHV_{ar} 16.2 MJ/kg) in the forest, costs of which are included in the final biomass price of 4.0 €/$GJ_{biomass}$. The dried chips are transported by truck to a BTL plant (loading costs 0.073 €/m^3; variable transport costs 0.08 €/ton/km; fixed transport costs 2.0 €/ton). The average transport distance to the BTL equals two-thirds of the area radius multiplied by 1.2 to accommodate for imperfectness of the existing road network.

On site of the BTL plant, the biomass is intermediary stored (one week capacity; 5.3 €/m^3 per year) before it is pre-treated (by torrefaction, 6.3.4.2) with 90–95 % efficiency, to yield a material that can be fed to the gasifier and allows stable gasification. The pre-treatment costs are fixed at 1.5 €/GJ of pre-treated material. In the oxygen-blown entrained flow gasifier the biomass is converted into biosyngas with 80 % chemical efficiency. The raw biosyngas is cooled, conditioned, and cleaned from the impurities. The on-specification biosyngas is used for FT synthesis to produce C_{5+} liquid fuels. Conversion efficiency from biosyngas to FT C_{5+} liquids is 71 %. All FT liquids products are equally considered as diesel fuel.

6.4.1 Capital Investment

Reliable cost data for Gas-to-Liquids projects are not available. Therefore, the investment costs for GtL as well as BtL plants in 2006 were derived from the off-the-record information on the EPC cost of the 34,000 bbld GTL plant built by Sasol-QP in Qatar.[44] From this information the total capital investments (TCI) are determined and with a constant scale factor of 0.7, the TCI is calculated for the whole scale range from 10 to 100,000 bbld. However, for smaller scales this results probably in an underestimate of the TCI costs as a smaller scale-factor would be more realistic, i.e. 0.6 or even 0.5 for 'real' small GTL plants. The scale dependency of the TCI of a GTL plant as well as of a BTL plant is presented in Figure 6.7.[44]

Based on assessment of the main equipments cost items of a BTL plant, it can be concluded that the TCI for a BTL plant is typically 60 % more expensive than a GTL plant with the same capacity, which is caused by the 50 % higher air separation unit (ASU) capacity, the 50 % more expensive gasifier due to the solids handling, and the requirement of a Rectisol unit for bulk gas cleaning. Although the approach followed is very simple, the results were in 2006 just as good, or probably even more accurate than

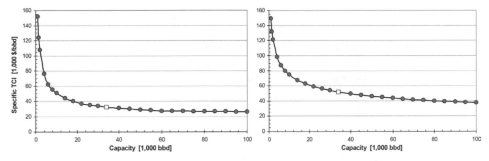

Figure 6.7 *Scale-dependency of specific TCI for GTL (left) and BTL (right) plants. Reproduced by permission of ECN.*

in-depth studies based on detailed assessment of equipment cost items. A recent trend though is that throughout the whole industry prices have gone up considerably.

6.4.2 Fischer-Tropsch Biodiesel Production Costs

Based on the capital costs for the BTL plant as presented in Figure 6.8, the annual capital (CAPEX) and operational (OPEX) costs are calculated with a depreciation period of 15 years (linear), a required IRR of 12 %, operation and maintenance (O&M) costs of 5 %, and a plant availability of 8000 h per year. Based on these assumptions as well as the assumptions on biomass (logistic) costs and conversion efficiencies, the costs of the produced FT liquids can be given as a function of the plant capacity (Figure 6.8).[44] It is clear that the costs for the conversion are the dominant cost factor at plant scales below 2000 MW$_{th}$ biomass input.

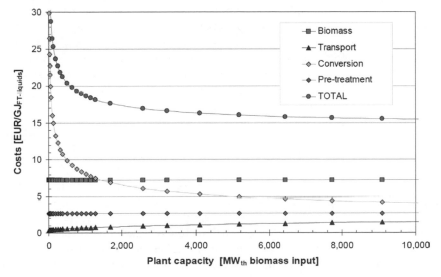

Figure 6.8 *Scale dependency of Fischer-Tropsch diesel fuel production costs. Reproduced by permission of ECN.*

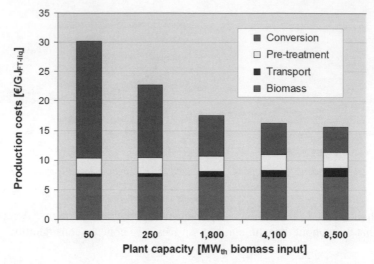

Figure 6.9   Scale dependency of FT-fuel production costs (15 €/GJ$_{FT}$ ≈ 55 €ct/L). Reproduced by permission of ECN.

6.4.2.1   Impact of Transport Distances

The transport, transhipment, and storage costs are only a small cost item, independent of the scale and related transport distances. The results also show that, in the assessed case, no advantage can be taken from decreasing the plant size, as the decrease in transport costs is completely outweighed by the increasing investment costs. In this assessment only land transport by truck is considered. In reference [45] various scenarios based on overseas biomass import are compared. In general can be stated that overseas transport would add approximately 1 €/GJ$_{FT}$ to the fuel costs.

6.4.2.2   Impact of Scale

In Figure 6.9, the cumulative FT-fuel production costs are shown in an alternative way, for five specific scales. The production costs decrease from 30 €/GJ$_{FT}$ for a 50 MW$_{th}$ plant to just above 15 €/GJ$_{FT}$ (i.e. approximately 55 €ct/l) at a scale of 8500 MW$_{th}$. The latter scale of the projected Shell Qatar plant is comparable to a conventional oil refinery. At large scale the biomass costs of 7.3 €/GJ$_{FT}$ make up half of the fuel costs. At small scale the investments costs are the determining cost item, i.e. two-thirds of the fuel costs.

6.4.2.3   Impact of Biomass Price

The costs of the biomass add 7.3 €/GJ$_{FT}$ to the FT-diesel fuel costs (independent of the scale). The 7.3 €/GJ$_{FT}$ follows from the overall biomass-to-fuel conversion efficiency of 56 %, i.e. for each GJ of FT fuel 1.8 GJ of biomass is required.[44] This illustrates the importance of systems with high biomass to fuel efficiencies. Operating a smaller BTL plant might be advantageous when cheap local biomass is available. In the case that biomass is available at 0.6 €/GJ$_{BM}$, FT-fuels can be produced at 15 €/GJ$_{FT}$ already in a 150 MW$_{th}$ biomass plant.[45] However, one can question how many of these locations will exist within a global biomass market. Therefore, based on economic considerations, it is advisable to direct technology development towards large BTL facilities. Additionally, it should be noted that the use of a scale factor of 0.7 for calculating the investment costs, most likely results in an underestimation of the costs for scales below 2,500 MW$_{th}$ (or 20,000 bbld). A factor of 0.6 for these smaller scales is probably more accurate, while a factor of 0.5 should be used for the even smaller scales below 5,000 bbld (600 MW$_{th}$).

6.5   Conclusions

To meet the longer-term market demands as mentioned e.g., in the Vision document of the European Technology Platform on Biofuels, and to gradually shift to biofuels with a better overall CO_2-reduction potential, the introduction of so-called second-generation biofuels is a necessity. Making H_2 and CO (syngas) from biomass is widely recognized as a necessary step in the production of second-generation biofuels. There are two major approaches in converting biomass into a syngas.

The first approach is based on fluidized bed gasification: biomass is converted to fuel gas at approximately 900 °C. This option requires almost no biomass pre-treatment, but the product gas needs downstream catalytic upgrading. A catalytic reactor is needed to reform the hydrocarbons to syngas. Support R&D is being performed at VTT (Finland), Värnamo (Sweden), and CUTEC (Germany). The fluidized bed gasification approach has the advantage that the gasification technology has already been developed and demonstrated with biomass for the production of heat and/or electricity.

The second approach is based on entrained flow gasification: fuel is converted at high temperature (1000–1300 °C) into a syngas with little or no methane and other hydrocarbons. The entrained flow gasification processes have already been developed and demonstrated on large-scale for coal. In some cases even mixtures of coal and specific biomass have been tested successfully. Most biomass feedstocks, however, are not suitable to be directly injected into an entrained flow gasifier because the fuel size needs to be small. Additional extensive pre-treatment (e.g. pyrolysis, torrefaction) is therefore required.

It should be realized that the above-mentioned options are not necessarily competing processes. The preference very much depends on boundary conditions, based on fuel type and fuel availability. Furthermore, the potential scale of the plants is an important issue. The back-end of the process generally needs to be as large as possible because of the dominant economy-of-scale effect in biofuel synthesis and upgrading. The front-end however involves biomass supply. This generally means that increasing plant size means higher feedstock costs because longer transport distances are involved.

There is however an attractive way to deal with this scale 'mismatch' by splitting the two parts: biomass is pre-treated in relatively small-scale plants close to the geographical origin of the biomass and the intermediate is transported to a central large-scale plant where it is converted into a biofuel. The pre-treatment should preferably result in a densified and easy to transport material. Conventional pelletization is an option, but more attractive is the use of dedicated pre-treatment that also produces a feedstock that can be used directly in the large-scale syngas plant, e.g. torrefaction or pyrolysis. Loss of efficiency due to pre-treatment then in general is compensated for by the logistic advantages of densification.

References

1. M.E. Dry, The Fischer-Tropsch Synthesis, in *Catalysis-Science and Technology*, J.R. Anderson; M. Boudart (Eds), Springer-Verlag, New York, USA, 159–255, 1981.
2. H. Boerrigter, H. den Uil and H.P. Calis, Green diesel from biomass via Fischer-Tropsch synthesis: new insights in gas cleaning and process design, in *Pyrolysis and Gasification of Biomass and Waste*, A.V. Bridgwater (Ed.), CPL Press, Newbury, UK, 371–383, 2003.
3. H. Boerrigter and H. den Uil, Green diesel from biomass by Fischer-Tropsch synthesis: new insights in gas cleaning and process design, RX–03-047, ECN, Petten, the Netherlands, 1–15, 2003.
4. (a) D. Hunt, Synfuels Handbook, Industrial Press Inc., McGraw-Hill Inc., New York, 1983 and (b) *Encyclopaedia of Science and Technology*, 7th ed., McGraw-Hill Inc., New York., USA, 1992.
5. E.D. Larson (2006) Advanced Technologies for Biomass Conversion to Energy, in *Proceedings of the 2nd Olle Lindström Symposium on Renewable Energy*, Bio-Energy, Royal Institute of Technology, Stockholm, Sweden, 2006.
6. (a) K. Hedden. A. Jess and T. Kuntze, From Natural Gas to Liquid Hydrocarbons, in *Edröl Erdgas Kohle, part 1*, **110**, 318–321, 1999, (b) Idem, *part 2*, **110**, 365–370, 1994, (c) Idem, *part 3*, **111**, 67–71, 1995 and (d) Idem, *part 4*, **113**, 531–540, 1997.
7. C. Higman and M. van der Burgt, Gasification, Elsevier Science, USA, 2003.
8. H. Boerrigter, A. van der Drift and R. van Ree, Biosyngas; markets, production technologies, and production concepts for biomass-based syngas, RX–04-013, ECN, Petten, the Netherlands, 1–37, 2004.
9. H.J. Veringa and H. Boerrigter, De syngas-route . . . van duurzame productie tot toepassingen, RX–04-014, ECN, Petten, the Netherlands, 2004.
10. (a) A. van der Drift, R. van Ree, H. Boerrigter and K. Hemmes, Bio-syngas: key intermediate for large scale production of green fuels and chemicals, in *proceedings of 2nd World Conference and Technology Exhibition on Biomass for Energy, Industry and Climate Protection*, **Vol. II**, 2155–2157, Rome, Italy, 2004 and (b) Idem, RX–04-048, ECN, Petten, the Netherlands, 1–4, 2004.
11. H. Boerrigter, A. van der Drift and E.P. Deurwaarder, Biomass-to-liquids: Opportunities & challenges within the perspectives of the EU Directives, in *5th Annual GTL (Gas-to-Liquids) Technology & Commercialization Conference & Exhibition*, 2006.
12. M. Kaltschmitt and H. Hartmann, Energie aus Biomasse, Grundlagen, Techniken und Verfahren, Springer-Verlag, Berlin, 770.
13. Conclusion from the Congress on Synthetic Biofuels – Technologies, Potentials, Prospects, Wolfsburg, Germany, 2004.
14. H. Boerrigter and A. van der Drift, Large-scale production of Fischer-Tropsch diesel from biomass: optimal gasification and gas cleaning systems., RX–04-119, ECN, Petten, The Netherlands, 2004.
15. H. Boerrigter, H.P. Calis, D.J. Slort, H. Bodenstaff, A.J. Kaandorp, H. den Uil and L.P.L.M. Rabou, Gas cleaning for integrated Biomass Gasification (BG) and Fischer-Tropsch (FT) systems, C–04-056, ECN, 2004.
16. M.A. Paisley, R.P. Overend, M.J. Welch and B.M. Igoe, FERCO's SilvaGas biomass gasification process commercialisation opportunities for power, fuels, and chemicals, in *The 2nd World Conference on Biomass for Energy, Industry, and Climate Protection*, Rome, Italy, 2004.
17. K. Whitty, State-of-the-art in black liquor gasification technology, in *IEA Annex XV meeting*, Piteå, Sweden, 2002.
18. A. van der Drift, An overview of innovative biomass gasification concepts, in *12th European Conference on Biomass for Energy*, Amsterdam, the Netherlands, 381-11384, 2002.

19. A. van der Drift, C.M. van der Meijden and H. Boerrigter, MILENA gasification technology for high efficient SNG production from biomass, *in 14th European Biomass Conference & Exhibition*, Paris, France, 2005.

20. EREC (European Renewable Energy Council): Renewable energy scenario to 2040, **16**, 2004.

21. L. Waldheim, Status of Chrisgas project; production of hydrogen-rich synthesis gas, in *Synbios, the syngas route to automotive biofuels*, Stockholm, Sweden, 2005.

22. E. Kurkela, Novel ultra-clean concepts of biomass gasification for liquid fuels, in *Synbios, the syngas route to automotive biofuels*, Stockholm, Sweden, 2005.

23. M. Claussen and S. Vodegel, The CUTEC concept to produce BtL-fuels for advanced power trains, in *International Freiberg Conference on IGCC and XtL technologies*, Freiberg, Germany, 2005.

24. A.G. Collot, Matching gasifiers to coals, ISBN 92-9029-380-2, IEA Clean Coal Centre, 63, 2002.

25. A. van der Drift, H. Boerrigter, B. Coda, M.K. Cieplik and K. Hemmes, Entrained flow gasification of biomass; ash behaviour, feeding issues, and system analyses, C–04-039, ECN, Petten, the Netherlands, 2004.

26. M. Schingnitz, Möglichkeiten zur Vergasung von Biomasse im Flugstrom, in: *Foerdergemeinschaft Oekologische Stoffverwertung e.V.*, Halle, Germany, 47–57, 2003.

27. B. Sander, G. Daradimos and H. Hirschfelder, Operating results of the BGL gasifier at the Schwarze Pumpe, in *Gasification Technologies*, San Francisco, USA, 2003.

28. M. Kanaar and C. Wolters, Fuel flexibility NUON Power Buggenum, in *Gasification, a versatile solution*, Brighton, UK, 2004.

29. F. Kamka, A. Jochmann and L. Picard, Development status of BGL gasification, in *International Freiberg Conference on IGCC and XtL technologies*, Freiberg, Germany, 2005.

30. R. Bain, Overview of US biomass gasification projects and fuel tax exemptions, in *Synbios, the syngas route to automotive biofuels*, Stockholm, Sweden, 2005.

31. A. van der Drift and H. Boerrigter, Synthesis gas from biomass for fuels and chemicals, C–06-001, ECN, Petten, the Netherlands, 2006.

32. P.C.A. Bergman, A.R. Boersma, R.W.R. Zwart and J.H.A. Kiel, Torrefaction for biomass co-firing in existing coal-fired power stations (BIOCOAL), C–05-013, ECN, Petten, the Netherlands, 1–72, 2005.

33. P.C.A. Bergman, A.R. Boersma, J.H.A. Kiel, M.J. Prins, K.J. Ptasinski and F.J.J.G. Janssen, Torrefaction for entrained flow gasification of biomass, C–05-067, ECN, Petten, the Netherlands, 1–51, 2005.

34. P.C.A. Bergman, Combined torrefaction and pelletisation – the TOP process, C–05-073, ECN, Petten, the Netherlands, 1–29, 2005.

35. A.V. Bridgwater, Fast pyrolysis of biomass: a handbook – Volume 2, CPL press, Newbury, UK, 2002.

36. E. Henrich, Clean syngas from biomass by pressurised entrained flow gasification of slurries from fast pyrolysis, in *Synbios, the syngas route to automotive biofuels*, Stockholm, Sweden, 2005.

37. E. Henrich and K. Raffelt, Two-stage rapid pyrolysis-entrained bed gasification of coal and solid wastes to synthesis gas, Patent EP1586621, 1–9, 2005.

38. E. Dinjus, E. Henrich, T. Kolb and L. Krebs, Synthesegas aus Biomasse, Verfahren des Fosrchungszentrums Kalrsruhe, in *Pyrolyse- und Vergasungsverfahren in der Energietechnik Bio-Fuel-Konzepte*, Freiberg, Germany, 2004.

39. M. Rudloff, Operation experiences of Carbo-V process for FTD production, in *Synbios, the syngas route to automotive biofuels*, Stockholm, Sweden, 2005.

40. E.D. Larson, Potential of biorefinery as large-scale production plant for liquid fuels in the forest and pulp industry, in *Synbios, the syngas route to automotive biofuels*, Stockholm, Sweden, 2005.

41. K. Sipilä, T. Mäkinen and P. McKeough, Raw material availability to synfuels production and remarks on RTD goals, results of the BioFuture project, in *Synbios, the syngas route to automotive biofuels,* Stockholm, Sweden, 2005.
42. I. Landälv, Status and potential of CHEMREC black liquor gasification, in *Synbios, the syngas route to automotive biofuels*, Stockholm, Sweden, 2005.
43. T. Ekbom, Techno-economics of biomass and black liquor gasification for automotive fuel production, in *Synbios, the syngas route to automotive biofuels*, Stockholm, Sweden, 2005.
44. H. Boerrigter, Economy of Biomass-to-Liquids (BTL) plants, C–06-019, ECN, Petten, the Netherlands, 1–29, 2006.
45. R.W.R. Zwart, H. Boerrigter and A. van der Drift, Integrated Fischer-Tropsch diesel production systems, Energy & Fuels, **20**, 2192–2197, 2006.

7

Plant Oil Biofuel: Rationale, Production and Application

Barnim Jeschke

*Co-founder and former Non-Executive Director, ELSBETT Technologies GmbH,
Munich, Germany*

7.1 Introduction

Of all biofuels, plant oil fuels have been around for the longest time. The former Rudolf Diesel already mentioned the use of plant oils for his engine designs in a 1912 patent application. Looking at the present global energy situation, plant oil fuel still seems to be one of the most promising biofuel options – with respect to economical, ecological and social benefits.

Plant oils work with diesel engines. As an important prerequisite, however, the underlying diesel engine technology needs to be adjusted in terms of heat management, fuel forwarding, filtration, fuel conditioning and fuel injection. Such specific 'plant oil technology' is necessary because of the different properties of plant oils compared to diesel fuel.

The legendary 'ELSBETT Engine', the first directly injecting car diesel engine worldwide, was already designed as a multi-fuel engine. With this engine, ELSBETT pioneered plant oil technology some 30 years ago, thereby initializing the modern biofuel movement.

Professional plant oil technology is globally applicable, with numerous oil fruits qualifying for practical applications in the respective regions. Suitable oil fruits include, for instance, rapeseed, sunflower, cannabis, and jatropha. With a worldwide energy scenario characterized by steadily increasing global energy shortages, political uncertainties and environmental problems, there is overwhelming evidence that the relevance of plant oil fuels will continue to grow at a rapid pace.

Biofuels Edited by Wim Soetaert and Erick J. Vandamme
© 2009 John Wiley & Sons, Ltd

7.2   Plant Oil Biofuels: the Underlying Idea

7.2.1   History of the Plant Oil Fuel Market

ELSBETT created the market for plant oil fuels by introducing the 'ELSBETT Engine', the first modern multi-fuel diesel engine worldwide. Some 30 years ago, Ludwig and Günter ELSBETT developed this engine – and thereby revolutionized the world of diesel engineering. This 3-cylinder engine was the first directly injecting, turbo charged car engine ever. At the same time, this engine was the first modern engine to run on plant oils. Developments as such laid the foundation for ensuing research and development work. This has resulted in the design of conversion kits which facilitate the operation of conventional diesel engines on plant oils.

Within the overall energy market, energy used for the operation of engines ('fuels') represents a major share. More than half of the worldwide fossil fuel supply is used to fuel the engines of around 800 million vehicles (Wüst, 2006).

Within the overall fuel market, fossil fuels (petrol and diesel fuels) predominate, with bunker oil being a by-product which is used only for big ship engines. At present, the overall biofuel share within the fuel market is still negligible, internationally averaging below 2.0 %.

It is a well-disseminated anecdote that in the first decade of the 20th century, Rudolf Diesel already experimented with peanut oil as a fuel for his engines. Later, biofuel options later gaining dubious relevance as fuel for military use, e.g. during the Second World War. Historic research of Hitler's first Eastern offensive suggests that his tanks running out of fossil fuel before reaching mineral oil sources in the Ukraine decisively weakened the German forces.

Besides such niches, biofuels were put on hold until the energy crises at the beginning of the Seventies. Since then, the market for plant oil fuels developed at an ever accelerating pace. From the year 2004 onwards, this growth has been fuelled by dramatically rising mineral oil prices and increasing political tensions in the Middle East. After being stuck in a market niche for more than two decades, plant oil applications are now considered a major fuel alternative for industrialized, emerging and developing countries.

The key market developments and milestones are compiled in Table 7.1.

7.2.2   Positioning of Plant Oils within the Biofuel Markets

The various present and future options within the biofuel market can be distinguished on the basis of the underlying motor technology and the corresponding approach to fuel supply.

Biofuels are specific to each engine type: Otto engines (originally designed for the use of petrol fuels), self-igniting diesel engines (originally designed for the use of diesel fuels), electrical engines (concepts, e.g. fuel cells are still in their infancy).

Generally, the use of diesel engines is increasing, especially for utility vehicles, but also for passenger cars. In Austria, more than 65 % of all newly registered cars have diesel engines. The application of plant oil fuels is based on the diesel engine concept.

Table 7.1   *Market milestones for plant oil fuels*

1978	Introduction of the 'Elsbett Engine', the first modern multi-fuel engine designed to run with both, Diesel and plant oil fuels.
1985	Beginning of technical conversions of Diesel engines for optional rapeseed oil operations, initially done only by Elsbett.
1996	Beginning of design and assembly of conversion kits to convert car Diesel engines for optional rapeseed oil operations; Elsbett among the first suppliers in this market; VWP as the first supplier of 'one-tank' conversion kits.
2001	Extending conversion kit solutions to heavy duty trucks and other utility vehicles; technology suppliers predominantly from Germany.
2003	Extending conversion solutions to modern Diesel engines with unit injection or common rail injection systems; Elsbett as the first supplier of common rail solutions.
2004	Rapidly increasing Diesel fuel prices in industrialized markets; exponential increase of demand for plant oil conversions; plant oils recognized and accepted as a viable alternative to fossil fuels at EU level.
2005	Rapidly evolving market demand especially in industrialized countries, accompanied by vivid political discussions and measures concerning the support of plant oil applications.
2006	Market teething problems mainly refer to 1) the use of unsuitable plant oils (e.g. soybean oil), 2) plant oil use/admixtures without (adequate) conversion, 3) negligent user maintenance.

Source: Elsbett AG.

Biofuels are either supplied in a blend with fossil fuels or as standalone fuels which require their own infrastructure:

- The blend approach is enjoying industry support of mineral fuel suppliers and dealers because it requires the least changes and investments on their part. Typical blends contain 2.0–5.75 % of biofuel.
- In contrast, the standalone approach is practically independent of fossil fuels and facilitates a decentralized, autonomous energy policy. Such an approach is most suitable for emerging and developing countries but also for other regions with good access to biofuels, e.g. plant oils, bio gas, or bioethanol. And for fully regenerative biofuels – such as plant oils – much greater benefits can be reaped if fossil fuels are substituted by up to 100 % with such biofuels.

While plant oils can be used in blends as well as stand-alone fuels, their most promising future lies in the latter.

7.3   Perspectives of the Plant Oil Fuel Market

7.3.1   The Market

Product-wise, the market for plant oil fuel includes all segments of the diesel engines market, specifically:

- mobile applications, such as passenger cars, trucks, agricultural vehicles, buses, trains, construction vehicles and other utility vehicles (e.g. fork lifts),
- stationary applications, such as generators and combined heat power systems.

Territory-wise, the market is of global reach, with varying motivations for the different regions:

* In industrialized countries, diesel fuel is expensive due to some high taxation via excise duties, which to the appeal of plant oil fuel.
* In rapidly emerging countries, energy shortages will increasingly limit the growth potential – unless alternative sources of energy are made available.
* Developing countries typically combine excellent conditions to grow suitable oil fruits (such as jatropha) with low labour costs. The possibility of farming oil fruits could even open doors to energy exporting ambitions and support agricultural activities in the poorest countries of the world, even in semi-arid areas.

Plant oil technology originates from Germany. But even there, the market is still in its infancy: In 2005, the plant oil quota of the total German fuel mix amounts to just 0.4 %. In the last two years, this market has at least doubled in Germany. In other countries, where this technology is still virtually unknown, growth rates can be expected to be significantly higher.

The rapidly growing market attracted numerous plant oil technology suppliers. Consequently, the competitive environment lost some of its transparency and – in some cases – credibility. C ompetition among plant oil technology suppliers can be distinguished according to technical capabilities and market focus:

* Suppliers of conversion solutions for 'older' diesel engine technology (chamber diesel engines and traditional directly injecting engines) are manifold and mainly located in Germany. Typically, suppliers within this segment are not engaged in research and development activities but follow 'me-too' approaches to convert outdated diesel engine technology. Since new vehicles nearly exclusively use more demanding common rail or unit injection systems, this 'me-too' segment will fade in the near future.
* Conversion solutions for 'modern' diesel engines – with either common rail or unit injection systems – are only provided by a handful of mid-sized companies which are mainly located in Germany: some of these already reach out to other countries via foreign sales partnerships. ELSBETT, for instance, has sales partnerships in a dozen countries.

7.3.2 Market Drivers

Plant oils are quite a suitable fuel for engine operations – without any relevant negative impact on engine performance and fuel consumption. However, the benefits of the plant oil option are much more fundamental than mere fuel cost savings.

Due to increasing supply shortages, fossil fuel prices are expected to rise steadily. Within this decade, current fossil fuel exporters such as Great Britain and Malaysia will follow countries like China and Indonesia, which already turned from fossil fuel exporters into import. In order to close this supply gap. the remaining exporting countries need to increase their exploration activities and to improve extraction technology. This cost-driving development shall be accompanied by a narrowing market supply and by persistent political risks and uncertainties, as exemplified by Iraq, which – according to its proven reserves – is ranked the fourth biggest fossil fuel exporter.

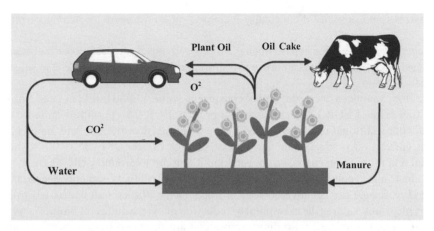

Figure 7.1   *Closed CO_2 loop for plant oil fuels.*
Source: Elsbett AG.

The world's fossil energy demand is currently met by a few fuel and gas exporters. Germany, for instance, imports 40 % of its gas from Russia, for Hungary this figure is 85 % (Follath, 2006). This entails political, military, and strategic risks. Current politics already demonstrate how far some mineral oil exporting nations use their resources for political power play.

Another major market driver for plant oil fuels are the related emission benefits. According to the Kyoto Protocol, the greenhouse effect has the deepest pollution impact on our environment. In this Protocol, more than 150 countries have committed themselves to increase their biofuel quota in order to reduce CO_2 emissions.

As displayed by Figure 7.1, plant oil fuel is part of a 'closed energy loop': oil fruits absorb as much CO_2 as are later emitted by the plant oil fuel. At the same time, one ton of plant oil roughly contains the same amount of energy as one ton of diesel fuel, but using plant oil avoids around 2.8 tons of CO_2 emissions. This opens new opportunities for trading emission certificates for countries that have ratified the Kyoto Protocol. In some parts of the world, e.g. in California, plant oil could also open markets where restrictive directives keep diesel fuel away from many applications.

Some desirable characteristics of plant oil fuels are:

- high calorific value: high energy density and high energy efficiency,
- lower soot emissions than diesel fuel, when burned,
- neither harmful nor toxic to humans, animals, soil or water,
- neither easily flammable nor explosive, and does not release toxic gases,
- easy to store, transport and handle,
- natural, recyclable form of energy that does not have to be chemically altered,
- no sulphur contained: it does not cause acid rain when used.

Matthew Simmons, former White House energy advisor, expects crude oil barrel prices to be between USD 200 and 250 for the 'next years' (Follath 2006). Such tendencies emphasize – and possibly dramatically increase – the cost-saving potential of plant oil fuels.

In recent years, the price gap between plant oils and diesel fuel widened so substantially, that diesel engine conversions amortize in a short period of time and then lead to substantial savings.

Take, for example, a heavy duty truck consuming some 40,000 litres per year. With a net diesel fuel price of EUR 1.15, a rapeseed oil price of EUR 0.85 per litre – and also taking into account additional expenses caused by the plant oil infrastructure and maintenance – a truck conversion produces ongoing annual savings of around EUR 10,000, while the material and installation costs of a conversion kit amount to around EUR 4000.

In industrialized countries, the price advantage of plant oil fuels is mainly due to taxation: the need to reduce the negative ecological impact of traffic emissions has led to public support measures, such as the exemption or reduction of excise duties for biofuels compared to fossil fuels. Complete excise duty exemptions are provided by a growing number of countries, including Germany, Belgium, and Austria. In other countries, the future tax policy towards biofuels is currently discussed, and related pilot projects are supported to gain first-hand experience. ELSBETT currently supports such projects in Ireland, Romania, France, and the Czech Republic.

For developing countries, plant oils also have good potential compared to fossil fuels: Here, fuel cost benefits are based on inexpensive, favourable conditions for plant oil production – contrasted by shortages in fossil fuel supply at a sufficient quantity and quality. Some key advantages of plant oil fuels for such regions are:

- ecological benefits include anti-soil erosion measures through oil fruit afforestation projects and organic fertilizer in the form of the oil cake derivate as a by-product;
- social benefits resulting from the support of local smallholder farmers and various employment effects at the levels of farming, oil mill operations, filling stations and retail, repair workshops, and counselling;
- political benefits in the form of an autonomous, self-sufficient energy policy and, possibly, an export option for technical plant oils;
- economical benefits due to fuel cost savings and the strengthening of regional monetary cycles: plant oils are often readily and cheaply available – while diesel fuel imports often represent a bottleneck.

7.4   System Requirements

A plant oil driven energy project typically includes the following system partners:

- oil fruit farmers,
- plant oil processors and plant oil dealers,
- plant oil storage and retailing,
- suppliers of plant oil conversion technology,
- repair workshops and mechanics for conversion and for after-sales services,
- optionally: local research institutes and local media.

Table 7.2   *Requirements and limit values for plant oil biofuels (DIN V 51605)*

Plant oil properties/contents	Unit	Limiting value	
		Minimum	Maximum
Density (at 15 °C)	kg/m³	900,0	930,0
Flash point (after Pensky-Martens)	°C	220	
Cinematic viscosity (at 40 °C)	mm²/s		36,0
Calorific value	kJ/kg	36.000	
Ignitability	DCN	39,0	
Carbon residue	% (m/m)		0,40
Iodine number	g Iod/100 g	95	125
Sulphur content	mg/kg		10
Overall contamination	mg/kg		24
Acid value	mg KOH/g		2,0
Oxidation stability (at 110 °C)	h	6,0	
Phosphorus content	mg/kg		12
Total content magnesia/calcium	mg/kg		20[e]
Ash content	% (m/m)		0,01
Water content	% (m/m)		0,075

Oil fruit farmers may include smallholder farmers as well as large cooperatives. This is due to the fact that plant oil production and processing does not require a minimum capacity scale. The choice of the respective oil fruit needs, of course, to consider technical feasibility.

Table 7.2 displays the most important properties for technical plant oils, as adopted from the currently concluded norming procedure on plant oils, DIN V 51605.

Considering the content of mono-saturated oleic acid, the low level of saturated fatty acid and the acceptable level of linoleic acid make rapeseed oil the ideal source of plant oil in Europe. Other sources widely used are palm oil, sunflower oil, soybean oil and jatropha which are economical and major sources of fuel in some parts of the world. However more studies on their application as biofuels are still needed. For instance, due to its low oxidation stability and other inherently deficient properties, soybean oil is not suitable for technical use.

Other criteria of the oil fruit choice relate to climatic conditions, crop yield, and overall agricultural concepts. The latter may, for instance, include a concept for recycling the resulting oil cake as organic fertilizer.

The production of technical plant oils can be achieved using two alternative methods: local, smaller-scale cold pressing facilities or through a refinery plant that works at larger, industrial scale:

• Cold pressing is carried out by smaller oil mills and provides a local, decentralized fuel supply. A cold pressing oil mill can run efficiently at relatively low capacity scales, with typical annual output of 500 – 10.000 tons. The incoming oil fruit seed is cleaned, pressed and filtered, accounting for the increasingly fine automotive fuel filters in order to prevent clogging-up. For this reason, we regard the maximum amount of acceptable contamination allowed by the Weihenstephan Standard (25 mg/kg) as slightly too high; values of below 15 mg/kg should be targeted instead.

- The production of fully refined plant oils requires an industrial scale, centralized processing plant. Although technical approaches for smaller refinery plants are known, efficient processing requires minimum annual capacities of 50.000 tons, with a more typical output of well above 100.000 tons. The fully refined plant oils are of homogeneous quality with very low levels of contamination.

In order to decide on the best production approach, some of the factors to be considered are:

- required annual quantity of plant oils within a certain region,
- required and given production knowledge,
- required investment amount to set-up or to adopt processing facilities.

Being a non-hazardous good, plant oil can be stored and transported without any restrictions. Storage should ideally avoid drastic temperature changes and direct exposure to UV light. In contrast to ordinary steel containers, plastic containers are permeable for UV radiation. Depending on the market and clientele, retailing may include a refuelling network of filling stations for small scale customers – or may refer to directly delivering larger amounts of oils to heavy users.

As a prerequisite for operating diesel engines on plant oils, some technical measures are necessary to convert existing engines. Suppliers of plant oil conversion technology should have specific expertise in both, the targeted engines and the sort of plant oil. A professional supplier should be able to offer customers a tailor-made solution by drawing on an already proven technology platform.

Mechanics are needed to actually install the conversion kit, to maintain the converted vehicle and to care about after-sales services, such as spare part handling. Such services are not rocket science and suppliers of plant oil technology should be willing and capable of providing self-explanatory installation manuals or personal training sessions.

An early involvement of local research institutes and selected media representatives will serve the quality, understanding and dissemination of plant oil projects.

7.5   Plant Oil Conversion Technology

Plant oil conversion technology compensates for some of the less desirable characteristics of plant oils:

- being a natural product, its quality varies considerably,
- it has a higher flash point than diesel (which is why you can cook with it!),
- it flows less freely than diesel (higher viscosity),
- it ignites less readily (lower cetane number).

Therefore, a technically sound adaptation of the fuel processing and the combustion process is needed, including the modification of the fuel transport, the enhancement of the fuel filtration, the heating of fuel and engine before and during operations, electronic control and, possibly, the modification of fuel injection. Consequently, a good conversion will account for the special characteristics of plant oil and should include:

- pre-warming of the fuel, fuel lines or motor,
- modifications to the fuel system/pumps,
- additional filter stages,
- adjustment of the electronic engine control,
- possible modification to the injection system,
- control elements and relays.

Existing conversion approaches can be categorized as either one or two tank systems. Regardless of the respective system, converted diesel engines can also run on diesel fuel or on plant oil/diesel fuel blends, if this is desired or required.

7.5.1   One Tank System

The one tank solution enables the user to completely waive diesel fuels. In order to make-up for the higher flash point of plant oil fuels, modifications of the injection system and/or the injection nozzles are necessary. Only below certain temperatures, typically around 0 °C, a blend with certain portions of diesel fuel will be required. If an auxiliary heating system or night-heater is desired, then a small secondary tank should be fitted for diesel, as these heaters do not run on plant oil.

ELSBETT is offering its one tank solution for all pre-chamber and swirl chamber engines, for traditional direct injecting engines and for selected unit injecting engines.

7.5.2   Two Tank System

The two tank system has a separate, smaller diesel tank to start and warm up the engine. As soon as the engine is warm, it switches over, automatically or manually, to take plant oil from the main, original tank of the vehicle for the remaining, main part of the journey. Before long stops where the engine might cool down, the driver switches back to diesel for the last few kilometres of the journey. This flushes the plant oil out of the injection system and prepares the engine for the next cold-start.

The auxiliary tank for a car typically has a capacity of 20–30 litres. In some countries the details of this change must be transcribed into the vehicle's registration by a testing centre, and it may also require a filler pipe on the outside of the vehicle. In this case, parts for external filling are provided with the ELSBETT conversion kit.

ELSBETT is providing semi-automated two tank systems for all unit injecting and common rail injecting engines, including modern heavy duty trucks.

7.6   The User Perspective

The market of plant oil technology, born with the ELSBETT Engine, is about as old as the personal computer: more than 25 years. However, market growth rates have increased only lately, indicating that the market is still in the phase of 'Early Adopters'.

The indicated early state of plant oil market penetration has the following characteristics:

- no industrial quality standard for plant oil technology,
- no cooperation of vehicle manufacturers, e.g. with respect to warranty,
- incomplete coverage with refuelling stations, especially for private users.

This is why the current customer profile is still that of an 'Early Adopter', with some pioneering spirit and some good understanding of the plant oil particulars. In order to enjoy the benefits of plant oil technology and fuel, such customers should have some knowledge concerning:

- engine suitability for plant oil conversion,
- choice of conversion technology and approach,
- operations and maintenance aspects.

7.6.1   Engine Suitability for Plant Oil Conversion

Plant oil technology is applied to diesel engines. Most vehicles with diesel engines regularly drive longer distances (so reach a good working temperature) and are fundamentally well suited for running on plant oil. However, to judge the suitability of a particular vehicle, and to avoid any disappointments in converting to plant oil, the following points should be considered:

- Technical suitability: Any vehicles fitted with a distributor-injection-pump made by CAV, Lucas, Stanadyne, RotoDiesel or Delphi are not suited for plant oil. Apart from these makes, all other injection systems are suitable, e.g. Common Rail-Systems from Delphi or Lucas.
- Condition of vehicle: The engine, injection system and electrical system of the vehicle must be in good condition. If that is not the case, for example, if the vehicle was previously run on (chemically aggressive) Biodiesel, the proper operation of a conversion cannot be guaranteed. In cases of doubt, the decompression and injection pressure should be checked.
- Journey profiles: Vehicles which do not regularly reach a good working temperature, e.g. as on longer journeys, should not be converted. On journeys of less than about 10 kilometres, the motor will not reach optimum thermal conditions. Besides reduced savings, there may also be a sharp increase in possible defects, such as mixing of plant oil in the motor oil, or wear and tear.
- Savings: The more fuel one uses, the greater the cost savings and environmental benefits of converting to plant oil.
- Filling facilities: The more often one can refill at home, the less one has to visit filling stations and fill up with diesel fuel (though diesel fuel can still be used).
- Personal attitude: A technically minded and inquisitive attitude towards the conversion will make it seem more straightforward.

The technology supplier of your choice will be able to inform you about the feasibility of converting specific engines to operate on plant oil. ELSBETT, for instance, is currently supplying conversion kits for over 2.400 vehicle types, including cars, trucks, and agricultural vehicles: A brief look into the online shop allows potential customers to determine if a vehicle qualifies for conversion.

7.6.2   Choice of Conversion Technology and Approach

Nowadays, the Internet greatly simplifies the selection of a plant oil conversion technology from competing offers. When searching for the 'best deal', the following questions might help in the decision-making process:

- Is the technology supplier directly involved in development work and have the offered solutions been sufficiently fleet-tested?
- Are the key components for the plant oil conversion kit specifically developed and produced for that very purpose?
- Is the supplier – often rather small companies – able to handle possible warranty claims? Are supplementing guarantee services offered?

7.6.3   Operations and Maintenance

Driving on plant oil will not cause the user any great inconvenience, but there are some guidelines which should be followed to avoid any problems:

1. Use shorter service intervals between lubricating oil changes. Alternatively, the engine oil can be analysed. These measures prevent potentially damaging 'polymerization' due to the mixing of plant oil with lubricating oil.
2. Add normal diesel fuel to the plant oil in the tank at sub-zero °C temperatures, according to the specifications of the respective plant oil technology supplier.
3. With two tank systems, run the last few kilometres on diesel fuel before letting the motor cool. This flushes the plant oil out of the injection system and prepares it for the next cold-start.

With further market penetration, plant oil biofuel applications will become more and more functional, convenient and self-explanatory for the user. And so will the related benefits.

References

Follath, Erich, *Der Treibstoff des Krieges – Im Kampf um Öl und Gas steuern die Großmächte USA und China auf eine gefährliche Konfrontation zu*, in: Der neue Kalte Krieg – Kampf um Rohstoffe, Erich Follath/Alexander Jung (Editors), Munich 21–52 (2006).
Wüst, Christian, *Bohrtürme zu Pflugscharen*, Spiegel Spezial 5/2006, 114–121 (2006).

8

Enzymatic Production of Biodiesel

Hideki Fukuda

*Division of Molecular Science, Graduate School of Science and Technology,
Kobe University, Japan*

8.1   Introduction

Alternative fuels for diesel engines are becoming increasingly important due to diminishing petroleum reserves and the environmental consequences of exhaust gases from petroleum-fuelled engines. A number of studies have shown that triglycerides hold promise as alternative diesel engine fuels.[1-8] However, the direct use of vegetable oils or oil blends is generally considered to be unsatisfactory and impractical for both direct-injection and indirect-type diesel engines. The high viscosity, acid composition, and free fatty acid content of such oils, gum formation due to oxidation and polymerization during storage and combustion, carbon deposits, and lubricating oil thickening are some of the more obvious problems.[9,10] Consequently, considerable effort has gone into developing vegetable oil derivatives that approximate the properties and performance of hydrocarbon-based diesel fuels. The problems encountered in substituting triglycerides for diesel fuels are mostly associated with their high viscosity, low volatility, and polyunsaturated character.[10]

Transesterification, also called alcoholysis, is the displacement of alcohol from an ester by another alcohol in a process similar to hydrolysis, except that an alcohol is employed instead of water. Suitable alcohols include methanol, ethanol, propanol, butanol, and amyl alcohol. Methanol and ethanol are utilized most frequently, especially methanol because of its low cost and its physical and chemical advantages. This process has been widely used to reduce the viscosity of triglycerides, thereby enhancing the physical properties of renewable fuels to improve engine performance.[11] Fatty acid methyl esters (MEs) obtained by transesterification can be used as an alternative fuel for diesel engines.

Biofuels   Edited by Wim Soetaert and Erick J. Vandamme
© 2009 John Wiley & Sons, Ltd

Among the attractive features of biodiesel fuel are (1) it is plant- not petroleum-derived and as such its combustion does not increase current net atmospheric levels of CO_2, a 'greenhouse' gas; (2) it can be domestically produced, offering the possibility of reducing petroleum imports; (3) it is biodegradable; and (4) relative to conventional diesel fuel, its combustion products have reduced levels of particulates, carbon monoxide, and, under some conditions, nitrogen oxides. It is well established that biodiesel affords a substantial reduction in SOx emissions and considerable reductions in CO, hydrocarbons, soot, and particulate matter (PM). There is a slight increase in NOx emissions, which can be positively influenced by delaying the injection timing in engines. [12–18]

Yamane et al. [12] reported that a biodiesel fuel with good ignitability, such as one with a high methyl oleate content, gives lower levels of NO, hydrocarbons, HCHO, CH_3CHO, and HCOOH, and also that soot formation is suppressed, since biodiesel is an oxygenated fuel having an O_2 mass fraction of 10 %.

Although chemical transesterification using an alkali-catalysis process gives high conversion levels of triglycerides to their corresponding MEs in short reaction times, the reaction has several drawbacks: it is energy-intensive, recovery of glycerol is difficult, the acidic or alkaline catalyst has to be removed from the product, alkaline wastewater requires treatment, and free fatty acids and water interfere with the reaction.

Enzymatic transesterification methods can overcome the problems of chemical transesterification mentioned above. In particular, it is notable that the by-product, glycerol, can be easily recovered without any complex process, and also that free fatty acids contained in waste oils and fats can be completely converted to MEs.

As a consequence of its advantages, there is considerable interest in exploring and developing the use of the enzymatic process for biodiesel fuel production. The present chapter describes the technologies relating to biodiesel fuel production by transesterification using lipase enzyme.

8.2   Enzymatic Transesterification by Lipase

The transesterification reaction with alcohol represented by the general equation shown in Figure 8.1a consists of a number of consecutive, reversible reactions as shown in

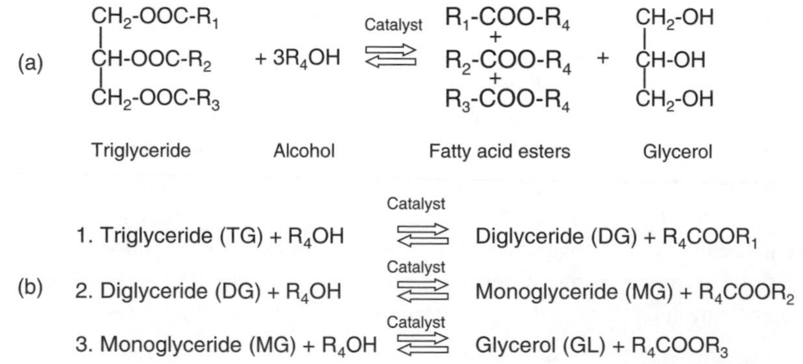

Figure 8.1   *Transesterification of triglyceride with alcohol. (a) General equation; (b) three consecutive and reversible reactions. R_1, R_2, R_3 and R_4 represent alkyl groups. (From Ref. 19, with permission of The Society for Biotechnology, Japan.)*

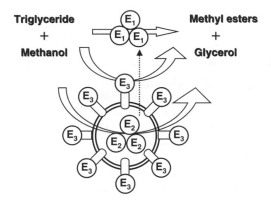

Figure 8.2   *Three types of lipase for methanolysis reaction. E_1: Extracellular lipase; E_2: Intracellular lipase; E_3: Cell-surface-displayed lipase.*

Figure 8.1b.[19] The first step is the conversion of triglycerides to diglycerides, which is followed by the conversion of diglycerides to monoglycerides and of monoglycerides to glycerol, yielding one fatty acid ester molecule from each glyceride at each step.[20,21]

The kinetics of triglyceride transesterification with methanol, i.e. methanolysis, catalyzed by *Rhizopus oryzae* lipase appears to follow a sequential reaction mechanism.[22] That is, triglycerides and partial glycerides are first hydrolyzed by lipase to partial glycerides and free fatty acids, respectively, after which methyl esters (MEs) are synthesized from free fatty acids and methanol (see Figure 8.1(b)). This suggests that, unlike in the case of alkali-catalyzed methanolysis, free fatty acids contained in used oils can be easily converted to MEs.

Three types of lipase, i.e. the extracellular, intracellular, and cell-surface-displayed lipases shown in Figure 8.2, can be utilized for methanolysis reaction. Since they can be prepared by different methods and also utilized in different forms, each reveals a range of characteristics in methanolysis reaction.

8.3   Use of Extracellular Lipases

Methanolysis reaction can be carried out using many kinds of extracellular lipase from different microorganisms. It is notable that both non-regiospecific and 1(3)-regiospecific lipases are utilized in methanolysis reaction and also that acyl migration in triglycerides is significant in the use of 1(3)-regiospecific lipase as it gives a higher conversion rate. The present section includes description of transesterification reactions using various types of extracellular lipase and a description of the effect of acyl migration on ME conversion with 1(3)-regiospecific lipase.

8.3.1   Transesterification with Various Types of Alcohol

As shown in Table 8.1 various types of alcohol – primary, secondary, and straight- and branched-chain – can be employed in transesterification using lipases as catalysts.[19] Linko

Table 8.1 *Enzymatic transesterification reaction using various types of alcohol and lipase*

Oil	Alcohol	Lipase	Conversion (%)	Solvent	Ref.
Rapeseed	2-ethyl-1-hexanol	C. RUGOSA	97	None	23
Mowrah, Mango, Kernel, Sal	C_4–$C_{18:1}$ alcohols	M. MIEHEI (*LIPOZYME IM-20*)	86.8–99.2	None	24
Sunflower	Ethanol	M. MIEHEI (*LYPOZYME*)	83	None	25
Fish	Ethanol	C. ANTARCTICA	100	None	26
Recycled restaurant grease	Ethanol	P. CEPACIA (*LIPASE PS-30*) + C. ANTARCTICA (*LIPASE SP435*)	85.4	None	27
Tallow, Soybean, Rapeseed	Primary alcohols[a]	M. MIEHEI (*LIPOZYME IM60*)	94.8–98.5	Hexane	28
	Secondary alcohols[b]	C. antarctica (SP435)	61.2–83.8	Hexane	
	Methanol	M. MIEHEI (*LIPOZYME IM60*)	19.4	None	
	Ethanol	M. miehei (Lipozyme IM60)	65.5	None	
Sunflower	Methanol	P. FLUORESCENS	3	None	29
	Methanol		79	Petroleum ether	
	Ethanol		82	None	
Palm kernel	Methanol	P. CEPACIA (*LIPASE PS-30*)	15	None	30
	Ethanol		72	None	

[a] Methanol, ethanol, propanol, butanol, and isobutanol.
[b] Isopropanol and 2-butanol.

et al.[23] have demonstrated the production of a variety of biodegradable esters and polyesters with lipase as the biocatalyst. In the transesterification of rapeseed oil with 2-ethyl-1-hexanol, 97 % conversion of esters was obtained using *Candida rugosa* lipase powder. De et al.[24] investigated the conversion of fatty alcohol esters (C_4–$C_{18:1}$) using immobilized *Mucor miehei* lipase (Lipozyme IM-20) in a solvent-free system. The percentage rate of molar conversion of all corresponding alcohol esters ranged from 86.8 to 99.2 %, while the slip melting points of the esters were found to increase steadily with increasing alcohol chain length (from C_4 to C_{18}) and to decline with the incorporation of unsaturation for the same chain length (as from C_{18} to $C_{18:1}$).

Transesterification of sunflower oil, fish oil, and grease with ethanol, i.e. ethanolysis, has also been studied. In each case, high yields of beyond 80 % were achieved using lipases from *M. miehei*,[25] *C. Antarctica*,[26] and *Pseudomonas cepacia*.[27]

Nelson et al.[28] investigated the ability of lipases in transesterification with short-chain alcohols to give alkyl esters. The lipase from *M. miehei* was the most efficient for converting triglycerides to their alkyl esters with primary alcohols, whereas that from *C. antarctica* was the most efficient for transesterifying triglycerides with secondary alcohols to give branched alkyl esters. Maximum conversions of 94.8–98.5 % for the primary alcohols methanol, ethanol, propanol, butanol, and isobutanol and of 61.2–83.8 % for the secondary alcohols isopropanol and 2-butanol were obtained in the presence of hexane as a solvent. In solvent-free reactions, however, yields with methanol and ethanol were lower than those obtained with hexane; in particular, the yield with methanol decreased to 19.4 %.

Mittelbach[29] reported transesterification using the primary alcohols methanol, ethanol, and 1-butanol, with and without petroleum ether as a solvent. Although the ester yields with ethanol and 1-butanol were relatively high, even in reactions without a solvent, only traces of ME were obtained with methanol. Abigor et al.[30] also found that, in the conversion of palm kernel oil to alkyl esters using *P. cepacia* lipase, ethanol gave the highest conversion rate of 72 %, while only 15 % (MEs) was obtained with methanol. Lipases are known to have a propensity to act better on long-chain than on short-chain fatty alcohols.[31,32] In general, the efficiency of the transesterification of triglycerides with methanol (methanolysis) is thus likely to be much lower than with ethanol in systems with or without a solvent.

8.3.2 Effective Methanolysis using Extracellular Lipase

Effective methanolysis reactions using extracellular lipase have been developed by several researchers and are summarized in Table 8.2.[19] Shimada et al.[33] found that immobilized *C. antarctica* lipase (Novozym 435) was the most effective for methanolysis of the lipases they tested. Since the enzyme was inactivated by shaking in a mixture containing more than 1.5 molar equivalents of methanol to the oil, they developed a method of adding methanol stepwise to avoid lipase inactivation. As a result, more than 95 % of the ester

Table 8.2 *Effective methanolysis processes with extracellular lipase*

Lipase	Regiospecificity	Process and operation	Methyl ester content (%)	Reaction time (h)	Ref.
C. antarctica (Novozym435)[a]	None	Repeated fed-batch operation[c]	96–98	48	33
		Continuous operation[d]	92–94	7[f]	34
		Fed-batch operation[c, e]	87	3.5	35
C. rugosa,[b] *P. cepacia,*[b] P. fluoresence[b]	None	Fed-batch operation[c]	80–100	80–90	36
R. ORYZAE (F-AP15)[b]	1(3)-regiospecific	Fed-batch operation[c]	80–90	70	22

[a] Extracellular lipase immobilized by ion-exchange resin.
[b] Extracellular lipase powder.
[c] Three-step addition of methanol: reaction mixture of oil/methanol (1:1, mol/mol) fed at each step.
[d] Continuous three-step flow reaction; a reaction mixture of oil/methanol (1:3, mol/mol) fed into each column.
[e] Pretreatment of immobilized lipase with methyl oleate and soybean oil.
[f] Overall residence time.

conversion was maintained even after 50 cycles of the reaction. In mixtures containing more than 1.5 molar equivalents of methanol, the excess amount of methanol remained as droplets dispersed in the oil. Because lipase may be inactivated when it contacts these methanol droplets, the authors stress that a reaction system in which methanol dissolves completely is necessary for the lipase-catalyzed methanolysis of oils. They also point out that the stepwise methanolysis procedure is suitable for adoption as an industrial process. Watanabe et al.[34] demonstrated effective methanolysis using two-step batch and three-step flow reaction systems with Novozym 435. The ME content in the final-step elute reached 90–93 %, and the lipase could be used for at least 100 days in both reaction systems without any significant decrease in the conversion rate. The effect of pretreatment of Novozym 435 on methanolysis for biodiesel fuel production was investigated by Samukawa et al.[35] Methanolysis progressed much faster when Novozym 435 was preincubated in methyl oleate for 0.5 h and subsequently in soybean oil for 12 h, with the ME content in the reaction mixture reaching over 97 % within 3.5 h under stepwise addition of 0.33 molar equivalents of methanol at 0.25–0.4 h intervals.

Kaieda et al. investigated the methanolysis of soybean oil with both non-regiospecific[36] and 1(3)-regiospecific[22] lipases in a water-containing system with no organic solvent. Of the non-regiospecific lipases, those from *C. rugosa*, *P. cepacia* and *P. fluorescens* displayed relatively high catalytic ability. In particular, the *P. cepacia* lipase yielded high ME content in a reaction mixture with as many as two or three molar equivalents of methanol to oil, which is attributed to the *P. cepacia* lipase having substantial methanol tolerance.[36]

The *R. oryzae* lipase, which exhibits 1(3)-regiospecificity (37–40), is also effective for the methanolysis of soybean oil. Kaieda *et al.*[22] investigated the effect of water content on ME conversion rate using the *R. oryzae* lipase. As shown in Figure 8.3, the lipase efficiently catalyzed methanolysis in the presence of 4–30 % water in the starting material, but the enzyme was nearly inactive in the absence of water. Despite of the use of 1(3)-regiospecificity lipase, the ME content in the reaction mixture reached 80–90 % under stepwise addition of methanol to the reaction mixture. This suggests that acyl migration from the *sn*-2 position to the *sn*-1 or *sn*-3 position in partial glycerides can occur spontaneously in a water-containing system. A high ME content of above 80 % was thus achieved. The details of acyl migration are described in the next section.

8.3.3   Phenomenon of Acyl Migration in Presence of Lipase

Oda et al.[41] analyzed the effect of water content and lipase on acyl migration of partial glycerides using an *R. oryzae* lipase solution. Thin-layer chromatography (TLC) analysis was carried out to confirm that acyl migration from the *sn*-2 to *sn*-1(3) position was promoted by *R. oryzae* lipase at different water contents (Figure 8.4a, b). Lane 1 shows methanolysate in 3 h reaction, Lane 2 heat-treated methanolysate, and Lanes 3–7 and Lanes 8–12 samples incubated at 30 °C with various amount of water for 2 h and 16 h, respectively.

In the absence of *R. oryzae* lipase, although 1(3)-MG (monoglyceride) content increased slightly with the passage of time, no difference in the progress of acyl migration was found to result from different water contents (Figure 8.4a). On the other hand, when samples with various amounts of water (Figure 8.4b) were incubated under the same conditions as in Figure 8.4a, but this time with addition of lipase, the content of 2-MGs and 1,2(2,3)-DGs

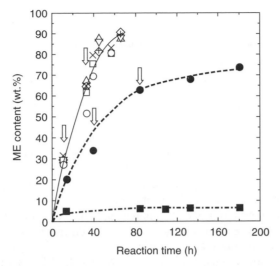

Figure 8.3 *Time courses of methyl ester content with 210 IU* Rhizopus oryzae *lipase added to enzyme solutions at the following concentrations (ml): closed square: 0.6; closed circle: 0.9; open triangle: 1.2; inverted open triangle: 1.5; open rhombus: 1.8; +: 2.4; x: 3.0; open square: 6.0; and open circle: 9.0. The reaction mixture, containing 28.95 g soybean oil, 0.6–9.0 ml distilled water (= 2–30 wt. % by weight of initial substrate), and 1.05 g methanol, was incubated at 35 °C with shaking at 150 oscillations/min in a bioshaker. Methanol (1.05 g) was added twice at the times indicated by arrows. (From Ref. 22, with permission of The Society for Biotechnology, Japan.)*

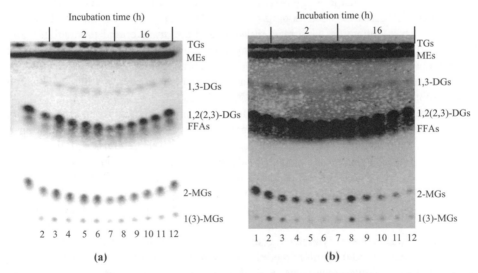

Figure 8.4 *Acyl migration of partial glycerides (a) without and (b) with* Rhizopus oryzae *lipase under the following water content conditions: 0 wt. % (Lanes 3 and 8); 5 wt. % (Lanes 4 and 9); 10 wt. % (Lanes 5 and 10); 15 wt. % (Lanes 6 and 11), and 30 wt. % (Lanes 7 and 12). Lane 1: initial 3 h methanolysate of soybean oil; Lane 2: heat-treated sample; Lanes 3–7: samples incubated for 2 h; Lanes 8–12: samples incubated for 16 h. (From Ref. 41, with permission of Elsevier.)*

(diglycerides) decreased with increasing water content and incubation time, indicating that *R. oryzae* lipase promotes acyl migration of partial glycerides from the *sn*-2 to *sn*-1(3) position. The free fatty acid content increased with increasing water content and incubation time, while that of other components (MEs and TGs) was not affected.

8.4   Use of Intracellular Lipase as Whole-Cell Biocatalyst

Methanolysis reaction can be carried out using extracellular or intracellular lipases, but extracellular lipases require purification by procedures that may be too complex for practical use. Furthermore, enzymes recovered through such operations are generally unstable and expensive. Consequently, there has been considerable research into the direct use of intracellular enzyme as a whole-cell biocatalyst.[42–44] The present section describes the whole-cell biocatalyst immobilization technique applied when utilizing *R. oryzae* intracellular lipase, the effect of cell membrane composition on methanolysis reaction, and the lipase localization within the cells.

8.4.1   Immobilization by BSP-Technology

To utilize the whole-cell biocatalyst in a convenient form, cells should be immobilized in such a way that they resemble the ordinary solid-phase catalysts used conventionally in synthetic chemical reactions. Of the many immobilization methods available, the technique using porous biomass support particles (BSPs) developed by Atkinson et al.[45] has several advantages over other methods in terms of industrial application: (1) no chemical additives are required, (2) there is no need for preproduction of cells, (3) aseptic handling of particles is unnecessary, (4) there are high rates of substrate mass transfer and production within BSPs, (5) the particles are reusable, (6) the particles are durable against mechanical shear, (7) bioreactor scale-up is easy, (8) costs are low compared to other methods. Because of its advantageous features, the BSP technique has been applied successfully in a wide variety of microbial, animal, and plant cell systems.[46]

In Figure 8.5, the extracellular and intracellular (whole-cell biocatalyst) lipase production processes are compared.[19] Unlike in the case of extracellular lipase, no purification or immobilization processes are needed in preparing whole-cell biocatalysts with BSPs, since immobilization can be achieved spontaneously during batch cultivation.

Utilizing *R. oryzae* cells immobilized within BSPs as whole-cell biocatalyst, Ban et al.[47] investigated the culture conditions for lipase production and the effects of cell pre-treatment and the water content of the reaction mixture on methanolysis in a 50-ml screw-cap bottle incubated on a reciprocal shaker. As shown in Figure 8.6, the *R. oryzae* cells were readily immobilized within the polyurethane foam BSPs during batch cultivation. Addition of olive oil or oleic acid to the culture medium as a substrate-related compound significantly benefited the intracellular lipase activity, while no glucose was necessary. As a result, when methanolysis was carried out with stepwise .addition of methanol using BSP-immobilized cells in the presence of 10–20 % water, the ME content of the reaction mixture reached 80–90 % with no organic solvent pretreatment.[47] This level of ME production is almost the same as that achieved using extracellular lipase.[22] Acyl migration can thus occur not only when extracellular lipase is used but also with *R. oryzae* cells as the whole-cell biocatalysts.

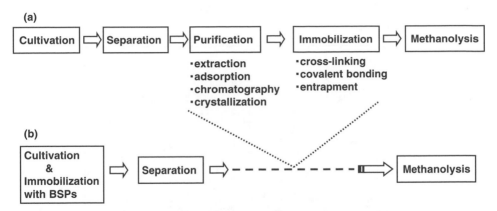

(a)

Cultivation ⇨ Separation ⇨ Purification ⇨ Immobilization ⇨ Methanolysis

Purification:
- extraction
- adsorption
- chromatography
- crystallization

Immobilization:
- cross-linking
- covalent bonding
- entrapment

(b)

Cultivation & Immobilization with BSPs ⇨ Separation ⇨ - - - - - - - - ⇨ Methanolysis

Figure 8.5 *Comparison of lipase production processes for methanolysis using (a) extracellular and (b) intracellular lipases. (From Ref. 19, with permission of The Society for Biotechnology, Japan.)*

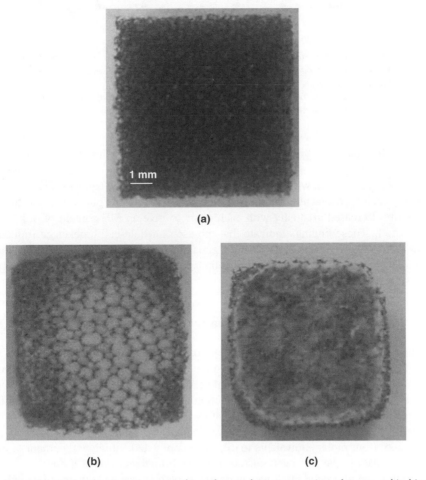

(a)

(b)   **(c)**

Figure 8.6 *Micrographs showing (a) empty, (b) surface and (c) cross-section of 6-mm cubic biomass support particle (voidage >97%; pore size: 50 pores per linear inch). Medium: polypeptone 70 g/l; $NaNO_3$ 1.0 g/l; KH_2PO_4 1.0 g/l; $MgSO_4 \cdot 7H_2O$ 0.5 g/l; olive oil 30g/l. Culture conditions: medium volume, 100 ml; temperature, 35°C; cultivation time, 80–90 h; shaking, 150 oscillations/min.*

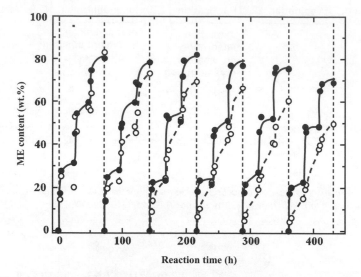

Figure 8.7   *Time course of methyl ester content in repeated methanolysis reaction using BSP-immobilized cells with and without glutaraldehyde treatment (0.1 % solution × 1 h at 25 °C). The procedure for immobilization within the BSPs was the same as described in Figure 8.6. Methanolysis was carried out for 6 batch cycles in the presence of 15 % water in the reaction mixture. Symbols: closed circle: treated cells; open circle: untreated cells. (From Ref. 48, with permission of Elsevier.)*

To stabilize the *R. oryzae* cells, cross-linking treatment with 0.1 % glutaraldehyde solution was examined.[48] As shown in Figure 8.7, the lipase activity of the cells thus obtained was maintained without significant decrease during six batch cycles, with the ME content in each cycle reaching 70–83 % within 72 h. Without the glutaraldehyde treatment, the activity decreased gradually with each cycle to give an ME content of only 50 % in the six batch. These findings indicate that the use of whole-cell biocatalyst immobilized within BSPs (47, 48) offers a promising means of biodiesel fuel production for industrial application because of the simplicity of the lipase production process as well as the stability of lipase activity over a long period.

8.4.2   Methanolysis in a Packed-bed Reactor using Cells Immobilized within BSPs

To achieve a higher ME content in the reaction mixture using cells immobilized within BSPs, Hama et al.[49] demonstrated effective methanolysis using a packed-bed reactor system shown in Figure 8.8. To increase the interfacial area between the reaction mixture and the cells immobilized within BSPs, the reaction mixture was emulsified by ultrasonication prior to reaction in each cycle. Figure 8.9 shows the time course of ME content during ten repeated-batch cycles of methanolysis in the packed-bed reactor. When stepwise addition of methanol (four molar equivalents to oils) was conducted, a high ME content of over 90 % was achieved in the first cycle of repeated-batch methanolysis at a flow rate of 25 l/h and a high value of around 80 % was maintained even after ten batch reaction cycles.

However, although a comparatively high ME content was obtained in the first cycle, the ME content decreased significantly during repeated methanolysis when using a screw-cap bottle under vigorous shaking. This is probably explained by cell exfoliation, as the

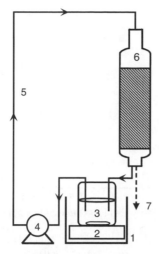

Figure 8.8 *Schematic diagram of packed-bed reactor. (1) Ultrasonic device, (2) magnetic stirring machine, (3) reservoir of substrates, (4) peristaltic pump, (5) Tygon tube, (6) glass column packed bed (25 mm$^\Phi$ × 400 mm^H) with whole cell biocatalysts, and (7) products.*

concentration of immobilized cells in the BSPs decreased significantly from 2.11 mg/BSP to 0.93 mg/BSP after the tenth cycle. These results suggest that a packed-bed reactor operated at an appropriate circulation flow rate offers a significant advantage in repeated methanolysis reaction by protecting cells from physical damage.

8.4.3   Effect of Fatty Acid Cell Membrane Composition

To stabilize the lipase activity of *R. oryzae* cells as whole-cell biocatalysts, Hama et al.[50] investigated the effect of cell membrane fatty acid composition on biodiesel fuel production. They found that oleic or linoleic acid-enriched cells showed higher initial methanolysis

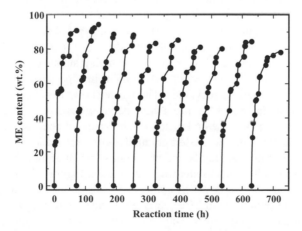

Figure 8.9 *Time course of methyl ester content during ten repeated cycles in packed-bed reactor.*

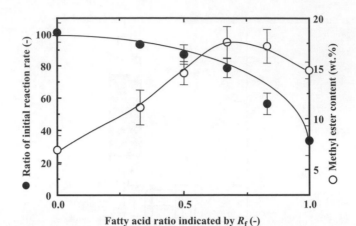

Figure 8.10   *Effect of fatty acid ratio (R_f) on development of initial reaction rate and methyl ester content. Left-hand scale shows ME content after 2.5-h in second batch cycle as percentage of that in first. (From Ref. 50, with permission of Elsevier.)*

activity than saturated fatty acid-enriched cells, among which palmitic acid-enriched cells exhibited significantly greater enzymatic stability than unsaturated fatty acid-enriched cells. It was assumed that fatty acids significantly affect the permeability and rigidity of the cell membrane, and that higher permeability and rigidity lead to increases in methanolysis activity and enzymatic stability, respectively.

Next, to investigate the optimal composition of the mixture of oleic and palmitic acid added to the culture medium, two cycles of repeated batch operation were carried out using mixtures with varying ratios of oleic to palmitic acid. Figure 8.10 shows ME content after 2.5 h in the first batch cycle and the initial reaction rate of ME production in the second batch cycle as a percentage of that in the first. This value reflects the degree of enzymatic stability: the higher it is the greater the enzymatic activity that can be expected.

Following increase in the fatty acid ratio, indicated by R_f [= oleic acid/(oleic acid + palmitic acid), w/w], there was no significant acceleration of the decrease in the initial reaction rate up to R_f value of 0.67, beyond which however it decreased rapidly. ME content showed its maximum value at R_f of 0.67, and decreased at R_f values greater or less than this critical level. It was thus concluded that there exists an optimal R_f value of 0.67 which yields both higher methanolysis activity and higher stability.

Figure 8.11 shows the time course of ME content in repeated methanolysis using *R. oryzae* cells obtained at various R_f values. The figure shows that, at R_f value of 1.0, both ME content and the initial rate of ME production decreased sharply with each cycle to give an ME content of only 5 % at the end of ten batch cycles. In contrast, at R_f value of 0.83 and 0.67, ME production rates decreased gradually with each cycle, and, in the latter case in particular, ME content was maintained at a high value of around 55 % even after ten batch cycles. These results are consistent with the data in Figure 8.10 and indicate that the optimal value of R_f in terms of both ME content and ME production rate is around 0.67.

Several studies have shown that free fatty acids affect the properties of the plasma membrane. Many have reported that membrane permeability is significantly affected by

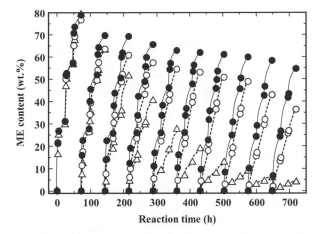

Figure 8.11 *Time course of methyl ester content during ten repeated batch cycles using cells cultivated at various R_f values. Symbols: (•) $R_f = 0.67$; (○) $R_f = 0.83$; and (△) $R_f = 1.0$. (From Ref. 50, with permission of Elsevier.)*

fatty acid composition in experiments using living cells[51] and lipid vesicle.[52–54] As for tolerance to ethanol[51] and toluene,[55] cell membrane rigidity appears to prevent their penetration. However, the effect of free fatty acids on the plasma membrane are still uncertain in terms of the mechanism of action and their location within the cell.

The findings outlined above indicate that BSP-immobilized cells are significantly stabilized by cultivation at optimal levels of R_f and could be used as whole-cell biocatalyst for practical biodiesel fuel production.

8.4.4 Lipase Localization in Cells Immobilized within BSPs

To determine the lipase localization in *R. oryzae* cells, Hama et al.[56] performed Western blot analysis. As can be seen in Figure 8.12, *R. oryzae* cells produced mainly two types of lipase with different molecular mass values of 34 and 31 kDa. Inside the cells, the 34-kDa lipase (ROL34) was bound to the cell wall and the 31-kDa lipase (ROL31) to the cell wall or membrane. Lipase in the cytoplasmic fraction was difficult to detect because of the small amount. It is notable that, in the suspension cells, the amount of membrane-bound lipase decreased sharply with cultivation time (Figure 8.12A), while in the immobilized cells large amounts of lipase remained even in the later period of cultivation (Figure 8.12B). Although the relationship between cell morphology and enzyme secretion depends on the fungal strain and the enzyme type, it was also found that cell immobilization strongly inhibited the secretion of ROL31 into the culture medium.

The N-terminal amino acid sequences of ROL34 and ROL31 were 'D-D-N-L-V' and 'S-D-G-G-K', respectively, clearly indicating the processing site of the lipase precursor (see Figure 8.13).[56] The *R. oryzae* lipase precursor consists of a signal sequence (26 amino acids), a prosequence (97 amino acids), and a mature region (269 amino acids), as deduced from the nucleotide sequence.[57] Of these regions, the prosequence contains a Lys–Arg

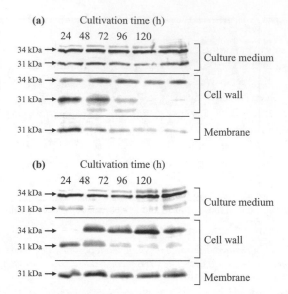

(a)   Cultivation time (h)

24 48 72 96 120

34 kDa →   ⎤
31 kDa →   ⎦ Culture medium

34 kDa →   ⎤
31 kDa →   ⎦ Cell wall

31 kDa →   ⎤ Membrane

(b)   Cultivation time (h)

24 48 72 96 120

34 kDa →   ⎤
31 kDa →   ⎦ Culture medium

34 kDa →   ⎤
31 kDa →   ⎦ Cell wall

31 kDa →   ⎤ Membrane

Figure 8.12 *Western blot analysis of Rhizopus oryzae lipase extracted from culture medium, cell wall, and membrane. Rhizopus oryzae cells were cultivated in (a) suspended and (b) immobilized cell cultures. Basal medium without oils and fatty acids was used for cultivation. Fifteen microliters of the lipase solution in each cellular fraction was subjected to SDS-PAGE electrophoresis. (From Ref. 56, with permission of The Society for Biotechnology, Japan.)*

sequence at amino acids −30 to −29 from its C-terminus, which is a kexin-like protease recognition site.[58] The N-terminal sequence of ROL34 was identical to the sequence of the precursor between residues 97 and 101, and that of ROL31 to the N-terminal 5-amino-acid sequence of the mature region. It was thus concluded that the processing at a C-terminal site of the Lys–Arg sequence produced ROL34, while ROL31 is the mature lipase given by cleavage of the N-terminal 28-amino-acid residue of ROL34.

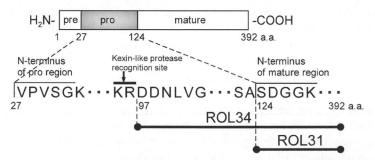

Figure 8.13 *Schematic representation showing precursor and processing site of ROL. After cleavage of the pre region, two patterns of processing occur in the pro region. First, the processing at a C-terminal site of the Lys–Arg sequence produces ROL34. Second, the complete cleavage of the pro region gives mature lipase with a molecular mass value of 31 kDa (ROL31). In the secretory pathway, ROL34 is localized in the cell wall and easily secreted into the culture medium, while ROL31 is tightly bound to the cell membrane. (From Ref. 56, with permission of The Society for Biotechnology, Japan.)*

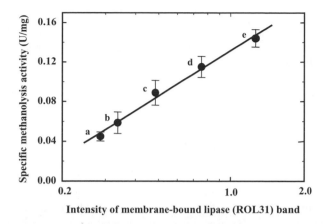

Figure 8.14   *Correlation between specific methanolysis activity and amount of membrane-bound lipase (ROL31). Letters in plots represent immobilized cells cultivated for 4 days (a) without oils and fatty acids, or with (b) stearic acid, (c) palmitic acid, (d) oleic acid, and (e) olive oil at 30 g/l each. The horizontal axis shows the sum of the intensities of the pixels inside the bands corresponding to each membrane-bound lipase detected by Western blot analysis. The specific methanolysis activity of the cells shows a linear relationship with the intensity of the membrane-bound lipase (ROL31) band in a semilogarithmic plot. (From Ref. 56, with permission of The Society for Biotechnology, Japan.)*

It is interesting that the N-terminal 28-amino-acid residue of ROL34 is critical for determination of the lipase localization. Generally, in eukaryotic cells, the topology of proteins in the membrane is determined by the translocation across the endoplasmic reticulum (ER) membrane after the cleavage of a signal peptide.[59] If once the secretory proteins are transported into the ER, they are located in the cell wall and finally secreted extracellularly. On the other hand, the proteins integrated into the ER membrane are finally accumulated in the cell membrane as it does to the ER membrane. Thus, the fact that ROL34 is localized mostly in the cell wall and very little in the membrane led to the hypothesis that the N-terminal 28-amino-acid residue of ROL34 plays an important role in the translocation of ROL across the ER membrane.

In the cell wall, where large amounts of ROL34 are localized, there was no significant difference according to the different substrate-related compounds added to the culture medium. In the membrane, however, cells cultivated with olive oil or oleic acid retained larger amounts of ROL31. As can be seen in Figure 8.14, there was significant correlation between the intracellular methanolysis activity and the amount of ROL31 localized in the membrane (56). These findings suggest that ROL31 localized in the membrane plays a crucial role in the methanolysis activity of *R. oryzae* cells.

8.5   Use of Cell-Surface Displaying Cells as Whole-Cell Biocatalyst

Cell-surface display of heterologous proteins using microorganisms has been widely utilized in various areas.[61,62] For instance, cell-surface display of enzymes such as glucoamylase and cellulase on bacteria[63–65], and yeast[66–68] has been found effective for the preparation of whole-cell biocatalyst. Antigenic proteins have also been displayed on the surface of yeast cells for the development of vaccines.[69–71] Of the various microorganisms available, yeast cells are suitable because of their safety, simplicity of genetic manipulation, and rigidity of cell-wall structure.[72]

The present section describes a novel cell-surface display system for the use of ROL and its application for methanolysis reaction.

8.5.1   Novel Cell-Surface Display System

In the most widely used yeast-based cell-surface display system, the gene encoding the target protein with the secretion signal is fused with the gene encoding the C-terminal half of α-agglutinin containing the putative glycosylphosphatidylinositol (GPI) anchor attachment signal sequence. However, the activity of enzymes whose active site is spatially near to the C-terminus[73] may be inhibited by fusion with an anchor protein, where *R. oryzae* lipase activity was strongly inhibited by fusion with a GPI anchor protein.[74]

For surface display of such enzymes as *R. oryzae* lipase in active form, Matsumoto et al.[75] have developed a new system based on the *FLO1* gene encoding a lectin-like cell-wall protein (Flo1p) in *Saccharomyces cerevisiae* (see Figure 8.15). Flo1p is composed of several domains: secretion signal, flocculation functional domain, GPI anchor attachment signal, and membrane-anchoring domain.[76–78] The Flo1p flocculation functional domain, thought to be located near the N-terminus, recognizes and adheres non-covalently to cell-wall components such as α-mannan carbohydrates, causing reversible aggregation of cells into flocs.[76,78,79] The new cell-surface display system shown in Figure 8.15 therefore consisted of the flocculation functional domain of Flo1p with a secretion signal and insertion site for the target protein. In this system, the N-terminus of target proteins such as ROL with pro-sequence (ProROL) are fused to the Flo1p flocculation functional domain.

To investigate the effect of the length of Flo1p on methanolysis reaction, the genes encoding the 1–1099 amino acids (*FS*) and 1–1417 amino acids (*FL*) of the flocculation functional domain of Flo1p were used.[75] For efficient expression of the fusion genes of *FSProROL and FLProROL* on the yeast cell-surface, the plasmids pWIFSProROL and pWIFLProROL were constructed as shown in Figure 8.15.

8.5.2   Flocculation Profile of Yeast Cells Displaying FSProROL and FLProROL Fusion Proteins

Figure 8.16 shows the flocculation ability of yeast cells displaying the FSProROL and FLProROL fusion proteins and wild-type non-flocculent yeast cells.[75] Interestingly, the yeast cells harboring pWIFLProROL at the late stage of cultivation seemed more flocculent than the wild-type flocculent strain *S. cerevisiae* ATCC60715 even though the FLProROL fusion protein produced had neither GPI anchor attachment site nor membrane-anchoring domain. (Figure 8.15). Meanwhile, yeast cells displaying FSProROL seemed to flocculate slightly in flask cultivation, although distinct ability was not detected in flocculation ability measurement. Surface display of the FL protein was just enough to obtain sufficient

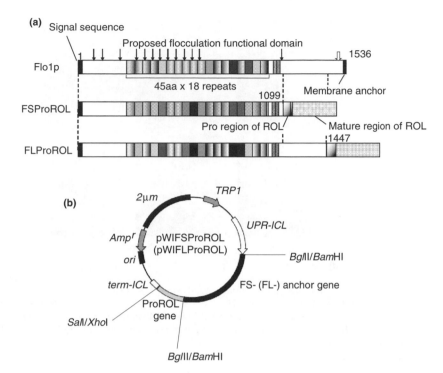

Figure 8.15 *(a) Structural features of Flo1p (76), FLProROL, and FSProROL. Arrowheads indicate possible N-glycosylation sites. Open arrow-head indicates possible GPI attachment site (amino acid 1514). Numbers written horizontally indicate length of proteins. The 45-amino-acid repeats comprise the proposed flocculation functional domain. (b) Yeast expression plasmids pWIFSProROL and pWIFLProROL. (From Ref. 75, with permission of American Society for Microbiology.)*

flocculation ability. This result leads to the conclusion that the hydrophobic C-terminal region of Flo1p (1448–1536 amino acids) is not necessary for flocculation, but that the C-terminal region of the FL anchor consisting of the 1100–1447 amino acids is.

8.5.3   Methanolysis Reaction Using Yeast Cells Displaying ProROL

In methanolysis reaction of plant oil, reaction substrates such as methanol and triglyceride can easily access surface-displayed lipase. Matsumoto et al.[75] therefore used yeast cells displaying FSProROL and FLProROL for biodesel production without permeabilizing treatment. Figure 8.17 shows the time course of ME content in methanolysis reaction. In yeast cells displaying FSProROL and FLProROL, ME content after 72-h reaction reached 78.6 % and 73.5 %, respectively, with two-step addition of methanol. Further, the initial reaction rate, defined as ME production rate, reached 3.2 g-ME/1εmin when 150 mg of lyophilized FSProROL-displaying cells were used (data not shown), a level as high as that for ROL in enzyme solution.[22]

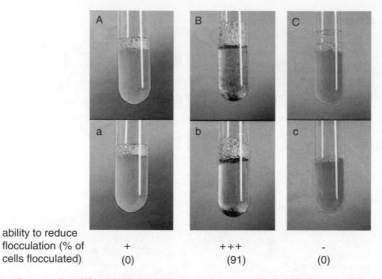

| | ability to reduce flocculation (% of cells flocculated) | + (0) | +++ (91) | - (0) |

Figure 8.16 Photographs of flocculating yeast during agitation (A, B and C) and after sedimentation for 5 s (a, b, and c). Panels: A and a: S. cerevisiae MT8-1/pWIFSProROL; panels B and b: S. cerevisiae MT8-1/pWIFLProROL; panels C and c: S. cerevisiae MT8-1/pWI3 (control). Ability of different constructs to trigger flocculation is indicated by +++ (full flocculation), + (slight flocculation), or – (no flocculation). Values in parentheses represent percentage of cells that settled after 1 min. (From Ref. 75, with permission of American Society for Microbiology.)

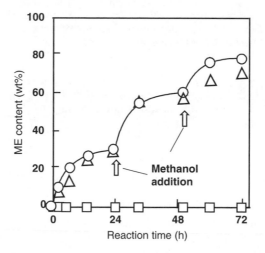

Figure 8.17 Time course of methanolysis reaction using yeast whole cells. ME content in reaction mixture is plotted against reaction time. ME content was defined as ME weight as a percentage of the total oil phase weight of the reaction mixture. Arrows show times of addition of 0.35 g methanol. Symbols: open circle: S. cerevisiae MT8-1/pWIFSProROL; open square: MT8-1/pWIFLProROL; open triangle: MT8-1/pWI3 (control). (From Ref. 75, with permission of American Society for Microbiology.)

With cell-surface-displayed ProROL, substrate molecules could easily access ProROL and no treatment was needed to catalyze methanolysis reaction. Since the initial reaction rate of FSProROL-displaying cells was as high as that of soluble ROL, the displayed FSProROL may have the same accessibility to substrates as free enzyme. For the industrial bioconversion process, lipases immobilized on the cell-surface are more cost effective and convenient, since these whole-cell biocatalysts can be prepared simply by cultivation and recovered easily.

8.6   Conclusions and Future Prospects

In recent years, biodiesel has become more attractive as an alternative fuel for diesel engines because of its environmental benefits and the fact that it is made from renewable resources. Used oils can also be utilized for making biodiesel fuel, thus helping to reduce the cost of wastewater treatment in sewerage systems and generally assisting in the recycling of resources.[80]

For the production of biodiesel fuel, an alkali-catalysis process has been established that gives high conversion levels of oils to MEs, and at present this is the method that is generally employed in actual biodiesel production. However, it has several drawbacks, including the difficulty of recycling glycerol and the need for either removal of the catalyst or wastewater treatment. In particular, several steps such as the evaporation of methanol, removal of saponified products, neutralization, and concentration, are needed to recover glycerol as a by-product.

To overcome these drawbacks, which may limit the availability of biodiesel fuel, enzymatic processes using lipase have recently been developed. Since the cost of lipase production is the main hurdle to commercialization of the lipase-catalyzed process, the use of intracellular lipase or cell-surface-displayed lipase as a whole-cell biocatalyst[47,48,75] is an effective way to lower the lipase production cost. Unlike in the case of extracellular lipase, these whole-cell biocatalysts can be prepared by simple cultivation and recovered easily.

However, to utilize these whole-cell biocatalysts for industrial application, a high ME content of 90–95 % in many repeated methanolysis reaction cycles is required. One potential solution is the use of a whole-cell biocatalyst possessing a non-specific lipase from a source such as *C. antarctica*[33] or *P. cepacia*[36] within the cell or on the cell-surface, since these lipases realize ME content of more than 95 %. Such a system could offer a promising prospect of realizing industrial biodiesel fuel production.

References

1. D. Bartholomew, Vegetable oil fuel, *J. Am. Oil Chem. Soc.*, **58**, 286A–288A (1981).
2. E. H. Pryde, Vegetable oil as diesel fuel: overview, *J. Am. Oil Chem. Soc.*, **60**, 1557–1558 (1983).
3. C. Adams, J. F. Peters, M. C. Rand, B. J. Schroer and M. C. Ziemke, Investigation of soybean oil as a diesel fuel extender: endurance tests, *J. Am. Oil Chem. Soc.*, **60**, 1574–1579 (1983).
4. C. L. Peterson, D. L. Auld and R. A. Korus, Winter rape oil fuel for diesel engines: recovery and utilization, *J. Am. Oil Chem. Soc.*, **60**, 1579–1587 (1983).

5. R. C. Strayer, J. A. Blake and W. K. Craig, Canola and high erucic rapeseed oil as substitutes for diesel fuel: preliminary tests, *J. Am. Oil Chem. Soc.*, **60**, 1587–1592 (1983).

6. C. R. Engler, L. A. Johnson, W. A. Lepori and C. M. Yarbrough, Effects of processing and chemical characteristics of plant oils on performance of an indirect-injection diesel engine, *J. Am. Oil Chem. Soc.*, **60**, 1592–1596 (1983).

7. E. G. Shay, Diesel fuel from vegetable oils: status and opportunities, *Biomass Bioenerg.*, **4**, 227–242 (1993).

8. M. Ziejewski and K. R. Kaufman, Laboratory endurance test of a sunflower oil blend in a diesel engine, *J. Am. Oil Chem. Soc.*, **60**, 1567–1573 (1983).

9. F. Ma and M. A. Hanna, Biodiesel production: a review, *Bioresour. Technol.*, **70**, 1–15 (1999).

10. A. Srivastava and R. Prasad, Triglycerides-based diesel fuels, *Renew. Sust. Energ. Rev.*, **4**, 111–133 (2000).

11. S. J. Clark, L. Wagner, M. D. Schrock and P. G. Piennaar, Methyl and ethyl soybean esters as renewable fuels for diesel engines, *J. Am. Oil Chem. Soc.*, **61**, 1632–1638 (1984).

12. K. Yamane, A. Ueta and Y. Shimamoto, Influence of physical and chemical properties of biodiesel fuel on injection, combustion and exhaust emission characteristics in a DI–CI engine, Proc. 5th Int. Symp. on Diagnostics and Modeling of Combustion in Internal Combustion Engines (COMODIA 2001), Nagoya, p. 402–409 (2001).

13. R. Varese and M. Varese, Methyl ester biodiesel: opportunity or necessity? *Inform*, **7**, 816–824 (1996).

14. W. Körbitz, The biodiesel market today and its future potential, *in* Martini, N. and Schell, J. S. (ed.), Plant Oils as Fuels. Springer–Verlag, Heidelberg (1998).

15. J. Sheehan, V. Camobreco, J. Duffield, M. Graboski, and H. Shapouri, An overview of biodiesel and petroleum diesel life cycles. Report of National Renewable Energy Laboratory (NREL) and US-Department of Energy (DOE). Task No. BF886002, May (1998).

16. A. Schäfer, Vegetable oil fatty acid methyl esters as alternative diesel fuels for commercial vehicle engines, *in* Martini, N. and Schell, J. S. (eds.), Plant oils as fuels. Springer–Verlag, Heidelberg (1998).

17. O. Syassen, Diesel engine technologies for raw and transesterified plant oils as fuels: Desired future qualities of the fuels, *in* Martini, N. and Schell, J. S. (eds.), Plant Oils as Fuels. Springer–Verlag, Heidelberg (1998).

18. T. Sams, Exhaust components of biofuels under real world engine conditions, *in* Martini, N. and Schell, J. S. (eds.), Plant oils as fuels. Springer–Verlag, Heidelberg (1998).

19. H. Fukuda, A. Kondo and H. Noda, *J. Biosci. Bioeng.*, **92**, 405–416 (2001).

20. B. Freedman, R. O. Butterfield, and E. H. Pryde, Transesterification kinetics of soybean oil, *J. Am. Oil Chem. Soc.*, **63**, 1375–1380 (1986).

21. H. Noureddini, and D. Zhu, Kinetics of transesterification of soybean oil, *J. Am. Oil Chem. Soc.*, **74**, 1457–1463 (1997).

22. M. Kaieda, T. Samukawa, T. Matsumoto, K. Ban, A. Kondo, Y. Shimada, H. Noda, F. Nomoto, K. Ohtsuka, E. Izumoto, and H. Fukuda, Biodiesel fuel production from plant oil catalyzed by *Rhizopus oryzae* lipase in a water-containing system without an organic solvent, *J. Biosci. Bioeng.*, **88**, 627–631 (1999).

23. Y.-Y. Linko, M. Lämsä, X. Wu, W. Uosukainen, J. Sappälä, and P. Linko, Biodegradable products by lipase biocatalysis, *J. Biotechnol.*, **66**, 41–50 (1998).

24. B. K. De, D. K. Bhattacharyya and C. Bandhu, Enzymatic synthesis of fatty alcohol esters by alcoholysis, *J. Am. Oil Chem. Soc.*, **76**, 451–453 (1999).

25. B. Selmi and D. Thomas, Immobilized lipase-catalyzed ethanolysis of sunflower oil in solvent-free medium, *J. Am. Oil Chem. Soc.*, **75**, 691–695 (1998).

26. H. Breivik, G. G. Haraldsson and B. Kristinsson, Preparation of highly purified concentrates of eicosapentaenoic acid and docosahexaenoic acid, *J. Am. Oil Chem. Soc.*, **74**, 1425–1429 (1997).

27. W. H. Wu, T. A. Foglia, W. N. Marmer and J. G. Phillips, Optimizing production of ethyl esters of grease using 95 % ethanol by response surface methodology, *J. Am. Oil Chem. Soc.*, **76**, 517–521 (1999).
28. L. A. Nelson, A. Foglia and W. N. Marmer, Lipase-catalyzed production of biodiesel, *J. Am. Oil Chem. Soc.*, **73**, 1191–1195 (1996).
29. M. Mittelbach, Lipase catalyzed alcoholysis of sunflower oil, *J. Am. Oil Chem. Soc.*, **67**, 168–170 (1990).
30. R. Abigor, P. Uadia, T. Foglia, M. Haas, K. Jones, E. Okpefa, J. Obibuzor and M. Bafor, Lipase-catalysed production of biodiesel fuel from some Nigerian lauric oils, *Biochem. Soc. Trans.*, **28**, 979–981 (2000).
31. Y. Shimada, A. Sugihara, H. Nakano, T. Kuramoto, T. Nagao, M. Gemba and Y. Tominaga, Purification of docosahexaenoic acid by selective esterification of fatty acids from tuna oil with *Rhizopus delemar* lipase, *J. Am. Oil Chem. Soc.*, **74**, 97–101 (1997).
32. Y. Shimada, A. Sugihara, Y. Minamigawa, K. Higashiyama, K. Akimoto, S. Fujikawa, S. Komemushi and Y. Tominaga, Enzymatic enrichment of arachidonic acid from *Mortierella* single-cell oil, *J. Am. Oil Chem. Soc.*, **75**, 1213–1217 (1998).
33. Y. Shimada, Y. Watanabe, T. Samukawa, A. Sugihara, H. Noda, H. Fukuda and Y. Tominaga, Conversion of vegetable oil to biodiesel using immobilized *Candida antarctica* lipase, *J. Am. Oil Chem. Soc.*, **76**, 789–793 (1999).
34. Y. Watanabe, Y. Shimada, A. Sugihara, H. Noda, H. Fukuda and Y. Tominaga, Continuous production of biodiesel fuel from vegetable oil using immobilized *Candida antarctica* lipase, *J. Am. Oil Chem. Soc.*, **77**, 355–360 (2000).
35. T. Samukawa, M. Kaieda, T. Matsumoto, K. Ban, A. Kondo, Y. Shimada, H. Noda and H. Fukuda, Pretreatment of immobilized *Candida antarctica* lipase for biodiesel fuel production from plant oil, *J. Biosci. Bioeng.*, **90**, 180–183 (2000).
36. M. Kaieda, T. Samukawa, A. Kondo and H. Fukuda, Effect of methanol and water contents on production of biodiesel fuel from plant oil catalyzed by various lipases in a solvent-free system, *J. Biosci. Bioeng.*, **91**, 12–15 (2001).
37. S. Okumura, M. Iwai and T. Tsujikawa, Positional specificities of four kinds of microbial lipases, *Agr. Biol. Chem.*, **40**, 655–660 (1976).
38. A. R. Macrae, Lipase-catalyzed interesterification of oils and fats, *J. Am. Oil Chem. Soc.*, **60**, 291–294 (1983).
39. M. Matori, T. Asahara and Y. Ota, Positional specificity of microbial lipases, *J. Ferment. Bioeng.*, **72**, 397–398 (1991).
40. H. Scheib, J. Pleiss, P. Stadler, A. Kovac, A. P. Potthoff, L. Haalck, F. Spener, F. Paltauf and R. D. Schmid, Rational design of *Rhizopus oryzae* lipase with modified stereoselectivity toward triacylglycerols, *Protein Eng.*, **11**, 675–682 (1998).
41. M. Oda, M. Kaieda, S. Hama, H. Yamaji, A. Kondo, E. Izumoto and H. Fukuda, Facilitatory effect of immobilized lipase-producing *Rhizopus oryzae* cells on acyl migration in biodiesel-fuel production, *Biochem. Eng. J.*, **23**, 45–51 (2005).
42. A. Kondo, Y. Liu, M. Furuta, Y. Fujita, T. Matsumoto and H. Fukuda, Preparation of high activity whole cell biocatalyst by permeabilization of recombinant flocculent yeast with alcohol, *Enzyme Microb. Technol.*, **27**, 806–811 (2000).
43. Y. Liu, H. Hama, Y. Fujita, A. Kondo, Y. Inoue, A. Kimura, and H. Fukuda, Production of *S*-lactoylglutathione by high activity whole cell biocatalysts prepared by permeabilization of recombinant *Saccharomyces cerevisiae* with alcohols, *Biotechnol. Bioeng.*, **64**, 54–60 (1999).
44. Y. Liu, Y. Fujita, A. Kondo and H. Fukuda, Preparation of high-activity whole cell biocatalysts by permeabilization of recombinant yeasts with alcohol, *J. Biosci. Bioeng.*, **89**, 554–558 (2000).
45. B. Atkinson, G. M. Black, P. J. S. Lewis and A. Pinches, Biological particles of given size, shape, and density for use in biological reactors, *Biotechnol. Bioeng.*, **21**, 193–200 (1979).

46. H. Fukuda, Immobilized microorganism bioreactors, *in* Bioreactor System Design, J. A. Asenjo and J. C. Merchuk (eds.), Marcel Dekker, New York, 1995.

47. K. Ban, M. Kaieda, T. Matsumoto, A. Kondo and H. Fukuda, Whole cell biocatalyst for biodiesel fuel production utilizing *Rhizopus oryzae* cells immobilized within biomass support particles, *Biochem. Eng., J.*, **8**, 39–43 (2001).

48. K. Ban, S. Hama, K. Nishizuka, M. Kaieda, T. Matsumoto, A. Kondo, H. Noda and H. Fukuda, Repeated use of whole-cell biocatalysts immobilized within biomass support particles for biodiesel fuel production, *J. Mol. Catal. B: Enz.*, **17**, 157–165 (2002).

49. S. Hama, H. Yamaji, T. Fukumizu, T. Numata, S. Tamalampudi, A. Kondo, H. Noda and H. Fukuda, Biodiesel-fuel production in a packed-bed reactor using lipase-producing *Rhizopus oryzae* cells immobilized within biomass support particles, *Biochem. Eng. J.* (to be submitted).

50. S. Hama, H. Yamaji, M. Kaieda, M. Oda, A. Kondo and H. Fukuda, Effect of fatty acid membrane composition on whole-cell biocatalysts for biodiesel-fuel production, *Biochem. Eng. J.*, **21**, 155–160 (2004).

51. H. Mizoguchi, S. Hara, Ethanol-induced alterations in lipid composition of *Saccharomyces cerevisiae* in the presence of exogenous fatty acid, *J. Ferment. Bioeng.*, **83**, 12–16 (1997).

52. H. Mizoguchi, S. Hara, Effect of fatty acid saturation in membrane lipid bilayers on simple diffusion in the presence of ethanol at high concentrations, *J. Ferment. Bioeng.*, **81**, 406–411 (1996).

53. M. Langner, S. W. Hui, Effect of free fatty acids on the permeability of 1,2-dimyristoyl-*sn*-glycero-3-phosphocholine bilayer at the main phase transition, *Biochim. Biophys. Acta*, **1463**, 439–447 (2000).

54. U. Locher, U. Leuschner, LUV's lipid composition modulates diffusion of bile acids, *Chem. Phys. Lipids*, **110** (2000), 165–171.

55. J. L. Ramos, E. Duque, J-J. Rodriguez-Herva, P. Godoy, A. Haïdour, F. Reyes, A. Fernandez-Barrero, Mechanism for solvent tolerance in bacteria, *J. Biol. Chem.*, **272**, 3887–3890 (1997).

56. S. Hama, S. Tamalampudi, T. Fukumizu, K. Miura, H. Yamaji, A. Kondo and H. Fukuda, Lipase localization in *Rhizopus oryzae* cells immobilized within biomass support particles for use as whole-cell biocatalysts in biodiesel-fuel production, *J. Biosci. Bioeng.*, **101**, 328–333 (2006).

57. H.-D. Beer, G. Wohlfahrt, R. D. Schmid and J. E. G. McCarthy, The folding and activity of the extracellular lipase of *Rhizopus oryzae* are modulated by a prosequence, *Biochem. J.*, **319**, 351–359 (1996).

58. M. Ueda, S. Takahashi, M. Washida, S. Shiraga and A. Tanaka, Expression of *Rhizopus oryzae* lipase gene in *Saccharomyces cerevisiae*, *J. Mol. Catal. B: Enz.*, **17**, 113–124 (2002).

59. R. Schülein, The early stages of the intracellular transport of membrane proteins: clinical and pharmacological implications, *Rev. Physiol. Biochem. Pharmacol.*, **151**, 45–91 (2004).

60. E. T. Boder and K. D. Wittrup, Yeast surface display for screening combinatorial polypeptide libraries, *Nat. Biotechnol.* **15**, 553–557 (1997).

61. G. Georgiou, H. L. Poetschke, C. Stathopoulos and J. A. Francisco, Practical applications of engineering gram-negative bacterial cell surfaces, *Biotechnol.*, **11**, 6–10 (1993).

62. S. Stahl and M. Uhlen, Bacterial surface display: trends and progress, *Trends Biotechnol.*, **15**, 185–192 (1996).

63. J. A. Francisco, C. F. Earhart and G. Georgiou, Transport and anchoring of beta-lactamase to the external surface of *Escherichia coli*, *Proc. Natl. Acad. Sci. U. S. A.*, **89**, 2713–2717 (1992).

64. J. A. Francisco, C. Stathopoulos, R. A. Warren, D. G. Kilburn and G. Georgiou, Specific adhesion and hydrolysis of cellulose by intact *Escherichia coli* expressing surface anchored cellulase or cellulose binding domains, *Biotechnology*, **11**, 491–495 (1993).

65. M. Little, P. Fuchs, F. Breitling and S. Dubel, Bacterial surface presentation of proteins and peptides: an alternative to phage technology? *Trends Biotechnol.*, **11**, 3–5 (1982).

66. T. Murai, M. Ueda, H. Atomi, Y. Shibasaki, N. Kamasawa, M. Osumi, T. Kawaguchi, M. Arai and A. Tanaka, Genetic immobilization of cellulase on the cell surface of *Saccharomyces cerevisiae*, *Appl. Microbiol. Biotechnol.*, **48**, 499–503 (1997).

67. T. Murai, M. Ueda, T. Kawaguchi, M. Arai and A. Tanaka, Assimilation of cellooligosaccharides by a cell surface-engineered yeast expressing beta-glucosidase and carboxymethylcellulase from *Aspergillus aculeatus*, *Appl. Environ. Microbiol.*, **64**, 4857–4861 (1998).

68. T. Murai, M. Ueda, M. Yamamura, H. Atomi, Y. Shibasaki, N. Kamasawa, M. Osumi, T. Amachi and A. Tanaka, Construction of a starch-utilizing yeast by cell surface engineering, *Appl. Environ. Microbiol.*, **63**, 1362–1366 (1997).

69. M. P. Schreuder, C. Deen, W. J. A. Boersma, P. H. Pouwels and F. M. Klis, Yeast expressing hepatitis B virus surface antigen determinants on its surface: implications for a possible oral vaccine, *Vaccine*, **14**, 383–388 (1996).

70. P. S. Schreuder, A. T. A. Mooren, H. Y. Toschka, C. T. Verrips and F. M. Klis, Immobilizing proteins on the surface of yeast cells, *Trends Biotechnol.*, **14**, 115–120 (1996).

71. R. I. Walker, New strategies for using mucosal vaccination to achieve more effective immunization, *Vaccine*, **12**, 387–400 (1994).

72. P. N. Lipke and R. Ovalle, Cell wall architecture in yeast: new structure and new challenges, *J. Bacteriol.*, **180**, 3735–3740 (1998).

73. H. D. Beer, G. Wohlfahrt, J. E. McCarthy, D. Schomburg and R. D. Schmid, Analysis of the catalytic mechanism of a fungal lipase using computer-aided design and structural mutants, *Protein Eng.*, **9**, 507–517 (1996).

74. M. Washida, S. Takahashi, M. Ueda and A. Tanaka, Spacer-mediated display of active lipase on the yeast cell surface, *Appl. Microbiol. Biotechnol.*, **48**, 499–503 (2001).

75. T. Matsumoto, H. Fukuda, M. Ueda, A. Tanaka and A. Kondo, Construction of yeast strains with high cell surface lipase activity by using novel display systems based on the Flo1p flocculation functional domain, *Appl. Environ. Microbiol.*, **68**, 4517–4522 (2002).

76. B. L. Miki, N. H. Poon, A. P. James and V. L. Seligy, Possible mechanism for flocculation interactions governed by gene *FLO1* in *Saccharomyces cerevisiae*, *J. Bacteriol.*, **150**, 878–889 (1982).

77. A. W. Teunissen, E. Holub, J. van der Hucht, J. A. van den Berg and H. Y. Steensma, Sequence of the open reading frame of the *FLO1* gene from *Saccharomyces cerevisiae*m, *Yeast*, **9**, 423–427 (1993).

78. J. Watari, Y. Takata, M. Ogawa, H. Sahara, S. Koshino, M. L. Onnela, U. Airaksinen, R. Jaatinen, M. Penttila and S. Keranen, Molecular cloning and analysis of the yeast flocculation gene *FLO1*, *Yeast*, **10**, 211–225 (1994).

79. M. Bony, D. Thines-Sempoux, P. Barre and B. Blondin, Localization and cell surface anchoring of the *Saccharomyces cerevisiae* flocculation protein Flo1p, *J. Bacteriol.*, **179**, 4929–4936 (1997).

80. T. Murayama, Evaluating vegetable oils as a diesel fuel. *Inform*, **5**, 1138–1145 (1994).

9

Production of Biodiesel from Waste Lipids

Roland Verhé and Christian V. Stevens

Faculty of Bioscience Engineering, Department of Organic Chemistry,
Ghent University, Ghent, Belgium

9.1 Introduction

The need for new energy resources is increasing continuously, not only because of the growing industrialization and population, but also due to the high mineral oil price, the shrinking fossil resources and the rather unstable political situation of the petroleum and natural gas producing regions in the world. Moreover, in some areas the nuclear energy supply is under quite severe political pressure. In addition the use of petroleum, coal and gas causes environmental concern due to the increasing emission of CO_2 and the greenhouse gas effects. Fuels from petroleum are major sources of air contaminants such as SO_2, CO and NO_x, particular matter, fine dust and volatile organic components. Apart from these observations the economical policy for agricultural production (e.g. CAP: Common Agricultural Policy in the EU) is leading to new innovations for the sustainability of the agro-industry.

All the factors mentioned above show the necessity for the use of alternative resources for energy. Biomass is certainly one of the sources of energy supply next to wind, solar and hydro-energy. The main advantage of the transformation of biomass into energy is the potential to reduce the CO_2 emission as the carbon source in biomass-derived energy is biogenic and renewable. In this way a nearly neutral CO_2-balance is obtained and not leading to additional greenhouse gases and global warming.

Biofuels Edited by Wim Soetaert and Erick J. Vandamme
© 2009 John Wiley & Sons, Ltd

One of the renewable resources for the production of 'green energy' are vegetable oils and animal fats. Although today the major part of lipids is still used for human food (80 %), animal feed (5 %) and industrial applications (15 %) it can be expected that the use for industrial purposes and for energy supply will be growing in the very near future.

The use of neat vegetable oils as transportation fuel has not been very successful in the past due to the high viscosity which is causing engine problems such as injector and filter coking and deposit formation. Expensive changes in diesel engines are necessary in order to use vegetable oils as fuel. However, the use of vegetable oils and/or animal fats, neat or as mixtures with diesel fuel, is a good alternative to be used in stationary diesel engines for the production of 'green electricity heat and power'. The best alternative in order to approximate the physical and chemical properties and the engine performance of petroleum based diesel is biodiesel. Biodiesel consists of esters of long chain fatty acids with short chain alcohols. Fatty acid methyl esters (FAME) are most often used as biodiesel due to the lower price of methanol in comparison with higher alcohols with the exception of the use of ethanol for biodiesel production in Brazil. For the automotive industry biodiesel is blended with the mineral diesel fuel.

Biofuels and biodiesel will not be able to solve the increasing energy demand. Apart from biofuels derived from natural resources, new alternative ways of converting whole crops have to be developed. Integral valorization in cogeneration plants are already producing electricity and steam from palm oil and waste lipid resources.

In addition there are major barriers in the commercialization of biodiesel from lipids. First, the primary use of vegetable oils will remain as a food ingredient. The increasing world population and the growing standard of living in developing countries is leading to a higher consumption of vegetable oils. The conflict between the use as human food/animal feed on one hand, and fuel use on the other hand, is already affecting the world vegetable oil market process.

The second obstacle for biodiesel commercialization are the high manufacturing costs which are mainly due to the higher costs of the raw materials, especially the virgin vegetable oils, but also to the cost of production. Both costs are dependent on the location, plant size, the value of the byproducts (especially the glycerine production) and the environmental conditions. In most cases, in order to reach the standards for biodiesel, the crude oils-fats are the subject of an intensive refining process which is an additional cost factor in the final biodiesel production.

At this moment the major raw materials for biodiesel production is food-grade canola oil in Europe, soybean oil in North-South America and palm oil in South-East Asia and in Africa.

Less expensive resources as raw material for biodiesel production are waste oils such as waste cooking and frying oils, rendered animal fat, recuperated oils and side-streams from the refining, next to the direct use of crude and non-refined vegetable oils and new non-edible resources such as yatropha oil and lipids derived from micro-algae.

The aim of this chapter is to provide a 'state-of-the-art' overview of the conversion of crude lipids, waste oils and alternative oil resources including the processing, additional refining and applications.

Books and reviews dealing with the production of biodiesel are available and furnish detailed information on the use of crude and refined oils and waste streams (Knothe, 2000;

Mittelbach and Remschmidt, 2004; Knothe et al., 2005; Inform, 2006; Zhang et al., 2003a, 2003b; Zheng et al., 2006; Kulkarni et al., 2006; Niederl and Narodoslawsky, 2006).

9.2   Alternative Resources for Biodiesel Production

Using cheap vegetable oils as feedstock, biodiesel is not able to compete economically with petroleum-based diesel fuel. This is mainly due to the high cost of the starting materials of which the costs can be estimated between 70 and 85 % of the total production costs. The result is that the overall production cost is higher than for the petroleum based fuel (Noordam and Withers, 1996).

Lower-cost sources of lipids can improve the economics of the biodiesel production. On the other hand, impurities and artefacts in these alternative resources and modifications in well-established production methodologies are sometimes minimizing the higher costs of using refined stocks. Nevertheless waste oils and other oil resources are much less expensive than virgin vegetable oils. Unrefined oils are normally 20–25 % cheaper than highly refined oils. Waste cooking and frying oils, rendered animal fat (tallow, lard, sheep, chicken, duck) are much less expensive than food grade vegetable oils.

Until now the majority of waste oils are commercially sold to the animal feed sector and are returning to the human food chain. Harmful compounds formed during the frying of oils and the presence of artefacts are therefore also entering the food chain. This policy has created the dioxin crisis in Belgium in 1999 due to the mixing of transformator oil containing PCB's and dioxins with animal fat, leading to a huge export blockage for a whole range of products.

In 2002 the European Union has banned the introduction of waste lipid streams in animal feed and therefore new applications have to be developed. The use of non-refined vegetable oils will involve new technologies and processes for the production of biodiesel. The production of biodiesel from new oil-bearing plants such as Yatropha, and Pongamia could lead to the recultivation of degraded land, waste land and deserts in India, Indonesia and Kenia. Acid oils from refining side streams, and wood extracts are other promising and cheap resources of biodiesel.

The use of these alternative resources is complicated by the presence of products which are normally absent in refined oils. The presence of water, free fatty acids, dimeric and polymeric materials and artefacts (agrochemical residues, contaminants, proteins) will give rise to additional processing and higher costs.

In this chapter the technology of the conversion of waste frying oils, recuperated vegetable oils and animal fats, non-refined vegetable oils and side streams from refining will be discussed.

9.3   Conversion of Waste Frying and Cooking Oils into Biodiesel

Alkaline catalysts (such as NaOH, KOH, NaOMe, KOMe), are widely used for the conversion of refined edible oils into biodiesel which meets the EU and US biodiesel standards. However, in major cases it will not be possible to transform waste oils and fats or crude non refined vegetable oils and animal fats into biodiesel with adequate standards.

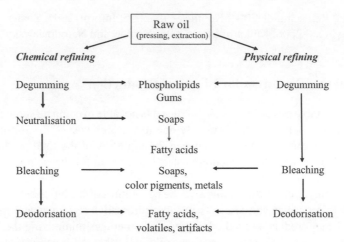

Figure 9.1   *Chemical and physical refining of vegetable oils.*

Refined oils have been the subject of an intensive additional pre-treatment in order to remove unwanted components. Crude lipids can be refined via two variations (Figure 9.1). The first process is chemical refining which involves a degumming step using water, phosphoric acid or citric acid in order to remove phospholipids and other gums. Neutralization with NaOH and KOH is transforming the free fatty acids into a soapstock which is removed by centrifugation followed by a bleaching process using an adsorbent (in most cases silicates) in order to remove high molecular colored materials (e.g. carotenoids, chlorophyll), artifacts (e.g. pesticide residue's, PCBs), traces of soaps, and metals (Fe, Cu). The last step involves a deodorization by heating the oil at 230–240 °C for 30–40 minutes in the presence of 1–2 % steam at 1–2 mm vacuum. During this step volatile compounds originating from the oxidation of the fatty acids (low molecular aldehydes, ketones, alcohols, acids) are removed as a deodorizer distillate next to the remaining acids and eventual contaminants (pesticide residues, PCB's, dioxins etc.). For some oils winterization (crystallization of high melting products such as waxes) is necessary in order to avoid the formation of deposits.

The second process, physical refining, does not include the neutralization of the FFA by base, in order to avoid the production of soapstock and the additional environmental cost for water purification. The degumming and bleaching steps are identical as in the chemical refining but in the deodorization step the free fatty acids are stripped off next to the other contaminants at a temperature of 240–260 °C for 40–60 minutes.

Vegetable oils and animal fats which have been refined using this sequence of treatments, can easily be converted in a two-step or continuous transesterification resulting in a biodiesel which fulfills the EU and US standards on condition that the suitable technology has been used. O'Brien et al. (2000) wrote an extensive overview of the refining of vegetable oils.

It will be advisable to use a similar technology for the treatment of waste oil as an initial step before transesterification, but this additional cost will make the use of waste oil less attractive. In addition, used oils contain more contaminants due to side reactions which occurred during the use of these oils.

In this section the physical and chemical reactions will be described which will be harmful for biodiesel production. In the next section the various technological approaches for the biodiesel production from waste fats and oils will be studied followed by a number of case studies.

9.3.1   Physical and Chemical Reactions in Lipids

Crude vegetable oils and animal fats are containing a number of artifacts which are changing the physical and chemical properties in comparison to refined oils. The presence of these unwanted components is due to a series of reactions: oxidation, hydrolysis, dimerization and polymerization. Similar reactions are occurring when refined oils and fats are used for food production, especially when their production is performed at higher temperatures, in oxidative conditions and in the presence of water. In addition, food product ingredients can migrate into the oil.

Physical changes in processed oils and fats are an increase in viscosity due to polymerization, an increase in the specific heat, changes in the surface tension and especially the higher tendency for foaming and browning and/or darkening of the color (Cvengros and Cvengrosova, 2004; Nawar, 1984).

9.3.1.1   Oxidation Reactions of Fatty Acids

Crude vegetable oils and fats and especially frying and cooking oils are oxidized due to the presence of unsaturated fatty acids. Oxidation can take place via three mechanisms.

Auto-oxidation, where oxygen is attacking the methylene function next to the double bond, involves an initiation, propagation and eventually a termination step. Auto-oxidation can only be stopped via the addition of anti-oxidants.

The second mechanism is photo-oxidation due to the presence of sensitizers (chlorophyll, haemoglobin, colorants, metals Fe, Cu) which are leading to radical formation followed by auto-oxidation or by the formation of singlet oxygen which is reacting 1200 times faster than triplet oxygen with unsaturated fatty acids.

An enzymatic oxidation involves the presence of lipoxygenase which is only active in the presence of a 1,4-pentadiene (e.g. linoleic, linolenic acid).

The rate of oxidation is dependent on the number of double bonds. The relative rate for oxidation in comparison to C18:1, is respectively 10 and 100 for C18:2 and C18:3.

The three oxidation pathways are giving rise to similar oxidation products. The primary reaction products are hydroperoxides and peroxides. The compounds undergo cleavage of the O-O bond with the formation of alkoxy radicals which are converted into secondary oxidation products such as aldehydes, ketones, alcohols and acids. Especially the formation of conjugated mono- and di-unsaturated aldehydes which are rapidly reacting with nucleophiles via a Michael addition can form unwanted components.

Considering the use of crude and waste lipids for the production of biodiesel it is important to consider the oxidative stability of the fatty acids. Therefore, used cooking and frying palm oil seem to be the best choice due to the higher degree of saturation followed by high oleic sunflower oil.

9.3.1.2   Hydrolysis

Hydrolysis of triglycerides is leading to the formation of free fatty acids, mono- and diglycerides and glycerol.

Distinction should be made between chemical hydrolysis especially at high temperature and in the presence of water and the enzymatic hydrolysis due to lipase activity present in the raw material. Lipases can be deactivated by sterilizing as in the palm oil production. Especially during cooking and frying, hydrolysis is occurring readily. During these processes there is a combination of oxidation and hydrolysis while also glycerol is converted into the highly reactive acrolein ($CH_2 = CH$-CHO).

A combination of these reactions and oxidative dimerization (see Section 9.3.1.3) is causing a multiplicity of chemical reactions in which undesirable compounds can be formed and which are not easy to remove from biodiesel (Guesta et al., 1993).

The quality of cooking and frying oils is expressed by the polar content. While the polar content of freshly refined oils is low (0.4–6.4 mg/100 g) successive frying gives rise to a multiplication by 100–1000. More unsaturation results in a higher polar content. The maximum amount of the polar content in the EU is set at 25 % in edible oil, otherwise the oil has to be discarded (Bastida, 2001).

9.3.1.3   Thermolytic Formation of Dimers and Polymers

Upon heating the triglycerides at high termperature (higher than 180 °C) in the absence of oxygen, the saturated fatty acids are converted to alkanes, alkenes, carbonyl compounds, CO and CO_2. The unsaturated fatty acids are forming linear dimers and dehydrodimers and polycyclic compounds. Especially the Diels-Alder reaction, which is occurring after conjugation due to isomerization of the non-conjugated systems, is leading to a dimer which can react further to trimers and polymers containing cyclohexene rings. Extensive heating is causing dehydrogenation with formation of aromatic rings leading to polyaromatic hydrocarbons (Nawar, 1984).

If the heating is performed in the presence of oxygen such as frying and cooking, dimerization is observed involving oxidation reactions leading to a variety of reactions such as ether linkage formation between unsaturated fatty acids and substitution of hydrogen by hydroxy functions.

9.3.1.4   Contamination of Waste Oils

Crude vegetable oils and animal fats can be contaminated during or after the production by proteins, carbohydrates, anorganic residues, packaging materials, cleaning agents, mineral oils and lubricants residues and dirt. The contaminants are preferably removed before biodiesel production by a combination of technologies mainly involving filtration.

However, these treatments are causing additional costs and should be taken into consideration for an economical production of biofuels in comparison with more refined feedstocks. Therefore it is important to know in advance the kind and the level of undesirable products.

9.3.2 Processing of Crude and Waste Lipids into Biodiesel

Due to the presence of water, excess of free fatty acids, the presence of phospholipids and various contaminants (polymers, residues of food products, packaging residues, anorganic compounds), the conversion of waste and crude lipids into biodiesel needs in many cases another technology and processing involving either a pre-treatment, transesterification combined with esterification or refining of the crude biodiesel.

9.3.2.1 Pretreatment of Waste Lipids

Highly contaminated waste lipids containing residues of food ingredients such as polar polymers, proteins, polycarbohydrates and anorganic materials should be purified either by filtration at 60–80 °C or by centrifugation. Especially animal fat from rendering (tallow, pork, chicken, duck) are containing high amounts of contaminants.

Prior to further treatment these raw materials are treated at 80 °C with 2 % of a phosphoric acid, citric acid or sulphuric acid solution followed by centrifugation (Verhé and Stevens, 2006; De Greyt and Kellens., 2006; Electrawinds, 2006). In this way the majority of the proteins, carbohydrates, anorganic residues (in the case of rendered tallow: residues of bones and skins) and partly the phospholipids are discarded.

Also steam injections (at 65 °C) and sedimentation have been used for waste oils. The results are a decrease in the moisture content, a reduction in the free fatty acid content, a substantial reduction in kinematic viscosity and unsaponifiable matter and an increase in the energy value. Especially the decrease of the water content and FFA results in a higher ester yield (Supple et al., 2002).

9.3.2.2 Removal of Water

The presence of water is affecting the alkaline-catalysed transesterification and the acid esterification either by saponification and/or by lowering the yields due to the reversibility of the reaction (Canackci et al., 1999). In many cases waste and crude lipids should be dried by heating the lipids at 70–90 °C under vacuum for at least 30 minutes before performing the transesterification or esterification.

9.3.2.3 Removal of Phospholipids

Crude and waste oil with a free fatty acid content lower than 3 % and phospholipids up to 300 ppm can be converted into biodiesel which is meeting the European standards without further treatment (Tomasevic and Silez-Marinkovic, 2003; Dorado et al., 2004).

However, phosphatides and gums can complicate the washing of crude biodiesel produced by transesterification due to the poor separation of the biodiesel-glycerol layer and the biodiesel-water layer resulting in a decrease of the ester yield.

Unrefined vegetable oils from which the phospholipids were removed are acceptable, they do not need bleaching nor deodorization and can be 10–15 % cheaper than highly refined oils (Kramer, 1995).

In many cases animal fats and palm oil do not contain sufficient amounts of phospholipids requiring degumming and can be used without further treatment with water and/or acid.

Crude and waste lipids containing high amounts of gums and phosphorous compounds should be treated by a degumming step involving washing the lipids with a solution of phosphoric and/or citric acid solution at 80 °C for 30 minutes. If necessary, adsorption on silica (Tonsil, Magnesol) can further reduce the gum content to acceptable levels in order to obtain a biodiesel with a P-content of less than 10 ppm.

During the degumming anorganic contaminants can be reduced, especially the concentration of alkali and earth alkali metals and the sulphated ash.

Application of the bleaching step resulting in removal of metals (Fe, Cu), colored materials, polar polymers, and soaps lowers the sulfur content (especially a problem in crude chicken fat).

9.3.2.4   *Removal of Free Fatty Acids*

It has been generally accepted that for producing biodiesel from crude and waste oils by transesterification with KOH in methanol, a free acid content up to 3 % does not affect the process negatively (Ahn et al., 1995). If higher amounts of free fatty acids are present, the reaction of the alkaline catalyst with the acids gives rise to soap formation which leads to loss of catalyst and in most cases to lower yields of esters due to the poor separation of the glycerol layer and the wash water layer in the presence of higher amounts of soap.

Free fatty acids can be removed from the feedstock by a variety of methods. As already mentioned steam injection results in a decrease of FFA in waste oils from 6.3 % to 4.3 % (Supple et al., 2002). Free fatty acids can also be removed by chemical neutralization with KOH or NaOH with formation of soapstock which is separated from the lipids by centrifugation. These separations are not always easy to perform and are resulting in lower yields. The soapstock can be used for the production of free fatty acids which in turn can be converted into biodiesel by acid esterification (see Sections 9.3.2.8 and 9.3.2.9).

An alternative removal of free fatty acids (Figure 9.2) is by extraction with the glycerol layer which has been separated from the ester layer during the alkali-based transesterification, since the glycerol layer contains a considerable amount of the alkaline catalyst in methanol. Mixing of this glycerol layer with the raw material containing free fatty acid results in neutralization and soap formation and the free fatty acids initially present are removed with the glycerol layer, which is separated from the oil. The separated oil layer can be used directly in the transesterification step (Harten, 2006).

On a laboratory scale, waste oils can be purified (reduction from 10.6 % FFA to 0.23 %) by passing through a column containing 50 % magnesium silicate and 50 % basic aluminium oxide (Ki-Teak et al., 2002).

Deacidification can be carried out by stripping off the free fatty acids in a simplified deodorizer at 200–240 °C and 1–4 mbar for 40 minutes in the presence of 1–2 % steam.

During this vacuum evaporation the free fatty acid content can be lowered from 20 % to 0.5 % in animal fat. Simultaneously oxidation products and partly sterols and tocopherols are removed as well as monoglycerides. In addition colored materials are destructed and contaminants (pesticide residues, polyaromatic hydrocarbons, PCBs dioxins etc) are discarded. However, cis-trans isomerization and dimer formation can occur which can give rise to higher viscosities and higher CP and CFPP (O'Brien et al., 2000).

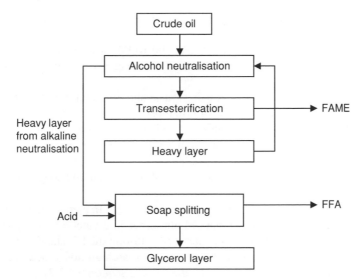

Figure 9.2 *Alcohol refining (Westfalia).*

9.3.2.5 Removal of Contaminants

The majority of the contaminants are mainly removed by either bleaching on adsorbens (clay, silicates, active carbon) and/or by deodorization.

Bleaching is carried out by treatment of the oil with 0.5–2 % bleaching earth at 90 °C for 30 minutes. During this process traces of soaps and phospholipids are removed next to residues of pesticides and artifacts, colored material and metals (O'Brien et al., 2000). Bleaching is an expensive process due to the high costs of the adsorbens, the disposal of the waste adsorbens (solid waste) and the loss of oil (oil can constitute 20–40 % of the weight of the disposed adsorbens).

Stripping of contaminants together with FFA is an alternative when the artefacts have a relatively low vapour pressure.

9.3.2.6 Production of Biodiesel from Crude and Waste Oil with Low Free Fatty Acid Content by Alkali-Transesterification

Alkali catalysts such as NaOH, KOH, NaOMe and KOMe are the most common catalysts for the transesterification of crude and waste lipids with a FFA content of less than 3 %. There is a direct link between lipid quality, expressed as the inverse of the FFA content, and the costs. For lipids with levels lower than 4 % FFA, the loss of the catalyst due to soap formation must be compensated by adding additional catalyst. For higher concentrations of FFA this will be impractical and uneconomical. An acid catalyst can also be used but acid transesterification is much slower (Canacki and Van Gerpen, 1999).

The choice of the catalyst is dependent upon the price, the work-up conditions and country. In the US normally NaOH is used while in the EU KOH is preferred due to a faster transesterification and the fact that the waste stream has a higher economical value due to the presence of K_2SO_4 which can be used as fertilizer. In general, sodium salts are cheaper than potassium salts.

Although they are more expensive, NaOMe and KOMe are also used, which have the advantage of greater safety, easier handling, better layer separation and purer glycerol (Markolwitz, 2004).

A typical transesterification reaction using waste and crude oil with less than 3 % FFA involves a six times molar excess of alcohol (in most cases methanol), 1–1.5 % (weight %) of catalyst at 20–65 °C for 30–90 minutes under vigorous stirring. As transesterification is a reversible reaction, the reaction is carried out in two steps, in many cases the first step yields a 75 % conversion into methyl esters.

The glycerol layer which contains residual alcohol, catalyst and soap is separated and fresh methanol and catalyst is added.

A two-step procedure gives normally a yield of 95–98 %. In a one-step procedure more catalyst is used. The reaction can be carried out either in a batch or continuous system. According to this procedure, biodiesel is obtained fulfilling the EU standards.

Several studies have been carried out by various research groups to determine the ratio oil/methanol/catalyst in order to produce biodiesel (see Section 9.3.2.7).

9.3.2.7 *Production of Biodiesel by Acid-Catalysed Interesterification*

The efficiency of the alkali-catalysed transesterification of waste oils is decreased in the presence of water and FFA mainly due to the soap – of and/or monoglycerides formation which is giving rise to poor separation of the ester and glycerol layers. Especially when the FFA content is higher than 1.5 %, acid-catalysed transesterification should be considered. On the other hand acid catalysis is much slower.

Acid catalysis involves the use of strong acids such as sulphuric acid and hydrogen chloride. The transesterification of vegetable oils using sulphuric acid has been studied by Canacki and Van Gerpen (1999). The most economical conditions involve the use of a molar ratio of 20:1 methanol: oil; 3 % weight of sulphuric acid at 60 ° for 48 hours.

The percentage conversion is increased with higher amounts of methanol and catalyst, and decreased dramatically in the presence of water with a conversion of 90 % for 0.5 % water to 32 % when 3 % water was present. The effect of the presence of FFA is far less dramatic: a level of 5 % FFA gives a 90 % conversion in comparison to a 75 % conversion at a 20 % FFA content.

A comparison has been made of transesterification with methanol in the presence of KOH and H_2SO_4 (Nye et al., 1983). The process using 0.1 % H_2SO_4 (65 °/40 h) and 0–4 % KOH (50 °/24 h) gives rise to an ester yield of 79.3 % and 91.9 % for a 3.6:1 methanol:oil ratio. However, the use of higher alcohols (ethanol-butanol) provided higher yields for the acid catalysed transesterification.

The reaction kinetics of the acid-catalysed transesterification of waste frying oil in excess of methanol to form FAME has been studied by Zheng et al. (2006). Rate of mixing, feed composition (molar ratio oil:methanol:acid) and temperature were independent variables. There was no significant difference in the yield of FAME when the rate of mixing was in the turbulent range 100–600 rpm. The oil:methanol:acid molar ratios and the temperature were the most significant factors affecting ester production. A pseudo-first-order reaction in the presence of a large excess of methanol, which drove the reaction to completion (99 % in 4 h) was observed using an oil:methanol:acid ratio of 1:245:38 at 70 °C and a ratio range of 1:75:1.9–1:245:3.8 at 80 °C. In these conditions also the FFA were rapidly converted into esters in only a few minutes. Also diglycerides present in the initial oil were rapidly transformed into FAME and very little monoglycerides were detected. For large-scale industrial productions this procedure seems to be less suitable due to the high excess of methanol which must be recovered.

Also waste palm oil has been transesterified in acid conditions (Al-Widyan and Al-Shyaukh, 2002).

Sulphuric acid and different concentrations of hydrogen chloride and ethanol at different levels of excess were used. Higher concentrations of catalyst (1.5–2.5 M) produced biodiesel in a much shorter time and of a lower specific gravity. Sulphuric acid was a much better catalyst than hydrogen chloride at 2.25 M.

Moreover, a 100 % excess of alcohol reduced the reaction time. The best process combination was the use of 2.25 M H_2SO_4 with 100 % excess of ethanol in a 3 h period.

Due to its low cost, H_2SO_4 seems to be the best catalyst for the acid transesterification of triglycerides in acid medium. The advantage is that simultaneously the FFA are converted into esters. However, when the amount of FFA is too high a quantitative ester formation can be prevented due to the water formation. A two-step reaction with water removal after the first step enables the formation of biodiesel with an acceptable level of FFA.

A calcium and barium acetate catalyst was developed for the production of biodiesel using feedstocks with high amounts of free fatty acids. A calcium and barium acetate catalyst was developed (Basu and Norris, 1996). However, the process is carried out at 200–220 °C and pressures of 2.76–4.14 MPa. In addition, the biodiesel produced contains too high levels of soaps and monoglycerides and the use of barium compounds is not environmentally friendly.

Another study compares the use of KOH and calcium and barium acetate (Rose and Norris, 2002). Using a methanol:oil ratio of 0.38, a mixture of 0.12 % barium acetate and 0.34 % calcium acetate at high temperature and high pressure during 2–3 hours reaction time resulted in similar esters yields (85–95 %) as 0.01 %–2 % KOH (methanol/oil ratio 0.2–0.28) for a reaction time of 1–2 h.

Other catalysts are acetates and stearates of calcium, barium, magnesium, mangane, cadmium, lead, zinc, cobalt and nickel (Di Serio et al., 2005). A ratio of oil:alcohol of 1:12 and a temperature of 200 °C for 200 min was used. Stearates gave better results due to the higher solubility than acetates in the lipophilic phase. These catalysts seem to be very promising due to their higher performance at lower catalyst concentration than Brönsted acids using lower alcohol to oil ratios and are less sensitive to the water content of the feedstock.

It can be concluded that homogeneous acid catalysts are suitable for the conversion of waste oils containing high levels of free fatty acids. However, the long reaction time, the high ratio of alcohol to oil, high concentrations of catalysts, separation of the ester layer and the extraction (washing) of the catalyst can jeopardize the economical conversion of waste oils into biodiesel.

A recent report (Lotero et al., 2005) reported the use of a solid acid catalyst for the conversion of high acidic oils into biodiesel and showed that these catalysts are performing a simultaneous esterification of FFA and transesterification of the tri-, di- and monoglycerides. In addition, they are less sensitive to the FFA content and the work-up (removal and re-use of the catalyst) is much easier.

9.3.2.8   *Two-Step Esterification-Transesterification for the Production of Biodiesel*

In order to produce a biodiesel from non-refined and waste oils and fats, a single acid or alkaline catalysis is not suitable to produce esters according to the biodiesel standards.

Many research teams have combined both acidic and alkaline catalysts in a two-step reaction in which the acid treatment converts the FFA into esters while the alkaline catalyst is performing the transesterification. Using both steps, biodiesel of high quality has been produced.

The dual process has been developed by Canacki and Van Gerpen (2003). A pilot plant hatch reactor (190 l) was built that can process high free fatty acid feedstocks using an acid-catalysed pre-treatment followed by an alkaline catalysed transesterification. In this way biodiesel from soybean oil, yellow grease with 9 % free fatty acids and brown grease with 40 % free fatty acid could be produced.

The effect of varying the reaction conditions is discussed together with the used technologies for separation, washing and refining.

The high free fatty acid feedstocks are treated with sulphuric acid in alcohol in order to reduce the free fatty acid content to less than 1 % by an acid esterification (Figure 9.3). The reaction mixture of fatty acid esters, mono-, di- and triglycerides were transesterified with methanol in the presence of KOH.

This two-step procedure is involving an acid-catalysed esterification followed by an alkaline transesterification which is a good technology to convert feedstocks with a high concentration of free fatty acids into biodiesel. The reaction rate is dependent upon the

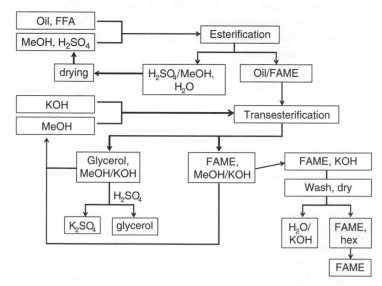

Figure 9.3   *Production of biofuels from recovered oil from frying oils.*

concentration of methanol and is increased by the amount of sulphuric acid. Oils with high concentrations of free fatty acids need a molar ratio of 40:1 of methanol/oil in comparison with a concentration of 6:1 for standard transesterification procedures. Also the presence of higher amounts of unsaponifiable material needs more methanol in comparison with a synthetic mixture of soybean oil with various amounts of palmitic acid. It is remarkable that substituting methanol by ethanol shows faster acidic esterification than using methanol.

Similar processes have been developed by Issariyakul et al. (2007) using methanol, ethanol or a mixture of both. The advantage of the mixture seems to be the more favourable equilibrium in the case of methanol in combination with the higher solvent properties of ethanol.

A highly efficient procedure in order to convert feedstocks with free fatty acid contents higher than 1 % and using short-chain alcohols and a combination of acidic esterification (sulphuric acid, alkylbenzene sulfonic acids) and a basic catalytic process at temperatures lower than 120 °C and under pressure (lower than 5 bar) was described by Lepper and Friesenhagen (1986) in the presence of ethylene glycol or glycerol. Due to the immiscibility with the oily phase, the water formed during the esterification can be entrained. The acid treated oil (which is containing fatty acid esters, probably mono- and diesters of ethylene glycol and glycerol) is treated with potassium hydroxide or sodium methoxide in a subsequent transesterification reaction. The residue of the glycerol layer can be reused for further reactions. The main advantage of this procedure is the shorter reaction time due the continuous removal of water in the first step and the elimination of soap formation during the alkaline step. The main disadvantages of the two-step acidic-alkaline reaction seems to be the removal of the catalyst in both steps involving two separations which can be time consuming, unless the acidic catalyst has been neutralized by an additional amount of the alkaline catalyst. Nevertheless it has been estimated by Canacki and Van Gerpen (1999) that savings in feedstock cost by using non refined lipids or waste lipids can result in an overall cost reduction of 25–40 % relative to refined resources.

Until now it is not clear if using a single acid catalysed esterification and transesterification is more economical than the two-step procedure (Zhang et al., 2003a). Therefore, the economic feasibilies of four continuous processes to produce biodiesel, including both alkali-and acid-catalysed processes, using waste cooking oil and the 'standard' process using refined edible oils were evaluated. The alkali-catalysed process using virgin vegetable oils has the lowest fixed capital costs, but a much higher raw material cost. The acid-catalysed process using waste cooking oil is less complex and is providing a lower total manufacturing cost, a more attractive after-tax rate return and a lower biodiesel breakeven price. However, the more corrosive nature of the acidic catalysts should be taken into consideration.

Plant capacity and especially prices of feedstocks were found to be the most significant factors affecting the economic viability of biodiesel manufacturing.

Other low-quality feedstocks, mainly consisting of free fatty acids, acylglycerides, unsaponifiable materials and contaminants such as soapstocks, deodorizer distillates and oils recovered from activated bleaching earth, can be converted mainly through a two-step process.

The soap stock obtained during the chemical neutralization of oils and fats is mainly composed of free fatty acids, acyl glycerides and phospholipids. It contains nearly 50 %

of water and due to the high pH, presence of polar lipids and water, the solid mixture has been disposed in landfills in the past and is creating environmental problems. Attempts have been made to convert the soap stock into biodiesel. Two alternatives have been used to transform soap stocks into biodiesel which is meeting the specifications.

The first process involves a complete hydrolysis by adding sodium hydroxide to the alkaline soap stock (2 h at 95 °C) resulting in a complete hydrolysis of the acylglycerides and phospholipids. After evaporation of the water, the free fatty acids are converted by a sulphuric acid catalysed esterification into the esters (Haas et al., 2000, 2001). The disadvantage of this process is the formation of sodium sulfate, which can precipitate during the reaction. An alternative procedure involves a process known as acidulation by introducing sulfuric acid and steam (Haas et al., 2003). The soaps are transformed into free fatty acids and two phases are separated; the upper oily layer known as acid oil and the lower aqueous layer. The formed sodium sulfate is removed through the aqueous layer and the upper layer is converted in biodiesel after drying by acid-catalysed esterification.

The second alternative consists in a first step of an esterification with methanol by acid catalysis, centrifuging the reaction product, removal of the acid-methanol layer and drying. The second step is a transesterification, washing with a water-methanol solution followed by centrifugation and removal of the water-methanol phase. Fractional vacuum distillation of the 90 % biodiesel gives a final product with a methyl ester content higher than 98 % (Marin et al., 2003).

Deodorizer distillate obtained during the last step in the physical refining of vegetable oils contains 70–90 % free fatty acids, mono- and diglycerides, sterols, tocopherols and eventually contaminants. It has been foreseen that the use of this fraction will be forbidden as an ingredient of animal feed due to the safety procedures in the agri-food chain.

A procedure has been developed to convert this feedstock into biodiesel via an acid esterification with sulfuric or p-toluene sulfonic acid/methanol (5 h – 65 °C) followed by transesterification according to standard procedures. In this way, a biodiesel can be obtained with an average of 90 % ester content. Vacuum distillation produces biodiesel of high purity leaving a residue of which the sterol fraction and/or the tocopherol fraction can be isolated through a fractional crystallization in methanol (Verhé and Stevens, 2006).

Used bleaching earth can contain up to 25 % of an oily fraction (up to 10 % free fatty acids). The recovered oil (by extraction with hexane) can also be a resource for biodiesel using the traditional two-step acidic-alkaline esterification and transesterification.

9.3.2.9   *Two-step Transesterification-Esterification for the Production of Biodiesel*

Raw vegetable oils, animal fats and waste oil from frying and cooking are converted into biodiesel via a new procedure developed by Ghent University (Verhé, 2007) involving as the first step a classical alkaline catalysed transesterification followed by an acid catalysed esterification. Both reactions are carried out in one reactor without separation of the layers between the two reactions. A requirement for the reaction is that the content of free fatty acids should be lower than 10 %, otherwise too much alkaline catalyst is deactivated by the fatty acids. The reaction is performed using the scheme in Figure 9.4, where the conversion of crude palm oil (CPO) into biodiesel is illustrated.

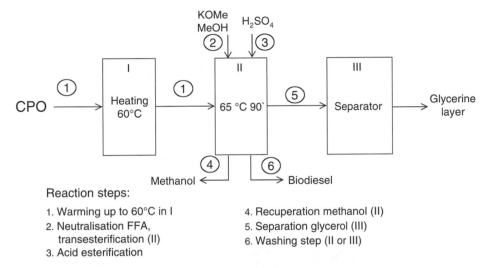

Figure 9.4 *Conversion of crude palm oil (CPO) by transesterification and esterification.*

Reaction conditions7

Step 1: Alkaline transesterification

catalyst: 0.6 % (w/w) NaOMe / 33 % in methanol + calculated amount of NaOMe in order to neutralize the FFA

methanol: 20 %

reaction: 65 °C / 1 hour

Step 2: acid catalysed esterification

catalyst: 2 % (w/w) H_2SO_4 + calculated amount to neutralize the excess of NaOMe and soaps reaction: 65 °C / 3 hours

The oil at 60 °C is mixed with the alkaline catalyst. During this transesterification the glycerol layer separates but is not removed. When the transesterification is finished, the calculated amount of H_2SO_4 is added and the mixture is stirred for 3 hours.

The methanol is evaporated, water is added and the glycerol/water layer is separated from the FAME layer. The FAME is washed with water until neutral. The biodiesel is dried under vaccuum at 65 °C.

The main advantage of this procedure is that the raw oils or fats do not need a pre-refining for the removal of phospholipids, gums, pigments and traces of proteins and carbohydrates. During the acidic esterification and the acidic work-up at 65 °C a degumming is taking place for the removal of the phosphorous containing compounds, gums and proteins while the pigments are decomposed and discarded with the aqueous layer.

In addition, the layer separation is much easier due to the fact that no emulsion is formed in acidic medium. Also traces of non-transesterified tri-, di- and monoglycerides are converted into FAME during the acidic transesterification which results in a higher ester content.

The properties of the biodiesel obtained with this procedure from crude palm oil, are shown in Table 9.1.

Table 9.1   *Properties of biodiesel from crude palm oil (CPO)*

Analysis	Result	EN limit
Ester content	96.8	Min 96.5 % (m/m)
Carbon Conradson residue	<0.001 %	Max 0.3 % (m/m)
Sulfated ash content	<0.01 %	Max 0.02 % (m/m)
Total contamination	44 mg/kg	Max 24 mg/kg
Acid value	1.7 mg/g	Max 0.5 mg KOH/g
Methanol content	0.02 %	Max 0.2 % (m/m)
Monoglyceride content	0.52 %	Max 0.8 % (m/m)
Diglyceride content	0.15 %	Max 0.2 % (m/m)
Triglyceride content	<0.01 %	Max 0.2 % (m/m)
Free glycerol	<0.01 %	Max 0.02 % (m/m)
Total glycerol content	0.15 %	Max 0.25 % (m/m)

9.3.2.10   Purification of Biodiesel

Recuperated waste oils from frying are containing a number of contaminants which cannot be removed by a pre-treatment and/or refining. The apolar and polar dimers, trimers and oligomers formed during heating and oxidation are not removed during bleaching and deodorization. The result is that dimeric, trimeric and oligomeric fatty acid methyl esters are present in the biodiesel lowering the FAME content (Norm 96,5 %) and sometimes causing precipitation.

The polymeric materials can be separated from the FAME by distillation of FAME at 200–220 °C/10 mbar. However, this is an additional costly step which increases the price of the biodiesel, but producing a biodiesel of excellent quality. The distillation residue can be used for the production of electricity by diesel engines or for fuel for steam production.

Additional purification of biodiesel can be performed by an adsorption process using magnesol.

9.4   Conclusion

Waste oils and crude vegetable oils and animal fats are an economical alternative for the production of biodiesel due to the lower price. Especially waste oils and fats which are not suitable any more to enter the feed-food chain are decreasing the conflict between food and non-food uses.

Excellent procedures have been developed for producing biodiesel meeting the majority of the international standards but in many cases additional refining, reaction steps and purification are necessary. The developed technologies are mainly batch processes due to more locally transport and production.

Blending with biodiesel produced from refined oils enhanced the possibilities to fulfil the international standards for biodiesel.

It can be foreseen that due to the constantly rising prices of the vegetable oils, used and waste oil streams will play an increasing role in the production of biodiesel.

References

Ahn, E., Koncar, M., Mittelbach, M., Marr, R. (1995). A Low-waste process for the production of biodiesel, *Sep. Sci. Technol.*, **30**, 2021–2033.

Al-Widyan, M., Al-Shyaukh, A. (2002). The experimental evaluation of the transesterification of waste palm oil into biodiesel, *Bioresource Technology*, **85**, 253–256.

Bastida, S., Sanchez-Muniz, F. (2001). Thermal oxidation of olive oil, sunflower oil and mix of both oils during forty discontinueous domestic fryings of different foods, *Food Sci. Technol. Int.*, **7**, 15–21.

Basu, H., Norris, M. (1996). Process of production of esters for use as a diesel fuel substitute using a non-alkaline catalyst. US Patent nr. 5, 525, 126.

Canacki, M., Van Gerpen, J. (1999). Biodiesel production via acid catalysis, *Trans ASAE*, **42**, 1203–1210.

Canakci, M., Van Gerpen, J. (2003). A pilot plant to produce biodiesel from high free fatty acid feedstocks, *Trans. ASAE*, **46**, 945–954.

Cvengros, J., Cvengrosova, Z. (2004). Used frying oils and fats and their utilisation in the production of methyl esters of higher fatty acids, *Biomass Bioenergy*, **27**, 173–181.

De Greyt, W., Kellens, M. (2006). De Smet-Ballestra Zaventem, Belgium, personal communication.

Di Serio, M., Tesser, R., Dimiccoli, M., Cammarota, F., Nasastasi, M., Santacesaria, E. (2005). Synthesis of biodiesel via homogeneous Lewis acid catalyst, *J. Mol. Catal.*, **239**, 111–115.

Dorado, M., Ballesteros, E., Mittelbach, M., Lopez, F. (2004). Parameters affecting the alkali-catalyzed transesterification process of used olive oil, *Energy Fuels*, **18**, 1457–1462.

Electrawinds (2006). Plassendale-Oostende, Belgium: personal communication.

Guesta, F., Sanchez-Muniz, Polonia-Garrido, S., Valera-Lopze, Arroyo, R. (1993). Thermoxidative and hydrolytic changes in sunflower oil used in frying with a fast turnover of fresh oils, *J. Am. Chem. Soc*, **70**, 1069–1073.

Haas, M., Bloomer, S., Scott, K. (2000). Simple high-efficiency synthesis of fatty acid methyl esters from soap stock, *J. Am. Oil Chem. Soc*, **77**, 373–379.

Haas, M., Michalski, P., Ruymon, S., Nunez, A., Scott, K. (2003). Production of FAME from acid oil, a by-product of vegetable oil refining, *J. Am. Oil Chem. Soc*, **80**, 97–102.

Haas, M., Scott, K., Alleman, T., McCormick, R. (2001). Engine performance of biodiesel fuel prepared from soybean soapstock: a high quality renewable fuel produced from a waste feedstock, *Energy Fuels*, **15**, 1207–1212.

Harten, B. (2006). Westfalia Separator. Practical Short Course on Biodiesel: Market Trends, Chemistry and Production. Istanbul Turkey, 13 August 2006.

Inform Supplement: Building Biodiesel, August 2006.

Issariyakul, T., Kulkarni, M.G., Dalai, A.K., Bakhishi, N.N. (2007). Production of biodiesel from waste fryer grease using mixed methanol /ethanol system. *Fuel Process Technol.*, **88**, 429–436.

Ki-Teak, L., Foglia, T., Chang, K. (2002). Production of alkyl ester as biodiesel from fractionated lard and restaurant grease, *J. Am. Oil Chem. Soc*, **79**, 191–195.

Knothe, G., Dunn R.O. (2000). Biofuels derived from vegetable oils and fats: in Gunstone F.D. and Hamilton R. J. Oleochemical Manufacture and Applications, CRC Press 106–163.

Knothe, G., Krahl, J., Van Gerpen, J. (2005). *The biodiesel Handbook*, AOCS Press Champaign, Illinois.

Kramer, W. (1995). The potential of biodiesel production, *Oils and Fats Int.*, **11**, 33–34.

Kulharni, M.G., Dalai, A.K. (2006). Waste Cooking Oil – An Economical Source for Biodiesel: a Review, *Ind. Eng. Chem. Res.*, **45**, 2901–2913.

Lepper, H., Friesenhagen, L. (1986). Process for the production of fatty acid esters of short-chain aliphatic alcohols from fats and/or oils containing free fatty acids, US patent no. 4, 608, 202.

Lotero, E., Liu, Y., Lopez, D., Suwannakarn, K., Bruce, D., Goodwin, J. (2005). Synthesis of biodiesel via acid catalysis. *Ind. Eng. Chem. Res*, **44**, 5353–5363.

Marin, J., Mateos, F., Mateos, P. (2003). *Grasas y aceites*, **54**, 130–137.

Markolwitz, M. (2004). Consider Europe's most popular catalyst, *Biodiesel Mag.*, **1**, 20–22.

Mittelbach, M., Remschmidt, C. (2004). *Biodiesel: The Comprehensive Handbook.* Martin Mittelbach, Graz, Austria, 330 p.

Nawar, W. (1984). Chemical Changes in lipids produced by thermal processing, *J. Chem. Ed.*, **61**, 299–303.

Niederl, A., Narodoslawsky, M. (2006). Ecological evaluation of processes based on by-products or waste from agriculture: Life cycle assessment of biodiesel from tallow and used vegetable oil. In: Feedstocks for the Future: Renewables for the Production of Chemicals and Materials, ACS Symposium, **921**, 239–252.

Noordam, M., Withers, R.V. (1996). Producing Biodiesel from Canola in the Inland Northwest: An Economic Feasibility Study, Idaho Agricultural Experiment Station, Bulletin no. 785, University of Idaho College of Agriculture, Moscow, Idaho.

Nye, M., Williamson, T., Desphande, S., Schrader, J., Snively, W., Yurkewich, T. (1983). Conversion of used frying oil to diesel fuel by transesterification: preliminary tests, *J. Am. Oil Chem. Soc*, **60**, 1598–1601.

O'Brien, R., Farr., W., Wan, P. (2000). *Introduction to fats and oils technology*, AOCS Press, p. 136–157.

Rose, P., Norris, M. (2002). Evaluate biodiesel made from waste fats and oils, Final report: agricultural Utilization Research Institute, Cookston, MN.

Supple, B., Holward-Hildige R., Gonzalez-Gomez, E., Leahy, J. (2002). The effect of steam treating waste cooking oil on the yield of methyl esters, *J. Am. Oil Chem. Soc.*, **79**, 175–178.

Tomasevic, A., Silez-Marinkovic, S. (2003). Methanolysis of used frying oil, *Fuel Process Technol.*, **81**, 1–6.

Verhé, R. (2007). Biodiesel production from waste streams. RRB3 Conference, 4-6 June 2007, Ghent, Belgium.

Verhé, R., Stevens, C. V. (2006). Non-published results.

Zhang, Y., Dube, M., McLean, D., Kates, M. (2003a). Biodiesel production from waste cooking oil: 2 Economic assessment and sensivity analysis. *Bioresource Technology*, **90**, 229–240.

Zhang, Y., Dube, M., McLean, D., Kates, M. (2003b). Biodiesel production from waste cooking oil: 1 Process design and technical assessment, *Bioresource Technology*, **89**, 1–16.

Zheng, S., Kates, M., Diebe, M.A., McLean, D.D. (2006). Acid-catalysed production of biodiesel from waste frying oil, *Biomass & Bioenergy*, **30**, 267–272.

10

Biomass Digestion to Methane in Agriculture: A Successful Pathway for the Energy Production and Waste Treatment Worldwide

P. Weiland

Bundesforschungsanstalt für Landwirtschaft, Institut für Technologie und Biosystemtechnik, Braunschweig, Germany

W. Verstraete

Faculty of Bioscience-engineering, Laboratory of Microbial Ecology and Technology, Ghent University, Ghent, Belgium

A. Van Haandel

Federal University of Paraíba, Department of Civil Engineering, Campina Grande, Brazil

10.1 Overview

Microbial conversion of energy crops and organic wastes has become one of the most attractive processes for energy production and waste treatment with resource recovery. It creates a wide range of positive environmental impacts because it reduces emission of greenhouse gases, improves the management of manure and organic wastes and reduces the demand for mineral fertilizers. Presently biogas is mainly used for electric power and

Biofuels Edited by Wim Soetaert and Erick J. Vandamme
© 2009 John Wiley & Sons, Ltd

heat production, but it can also be applied as an automotive fuel or for the production of hydrogen, which can be used in fuel cells. Biogas production in the agricultural sector is a very fast growing market, especially in many European countries. This chapter presents some aspects of the current situation in Germany and Brazil. The first has the highest number of agricultural biogas plants in Europe. The second has a large potential of biogas production from the residues of the bioalcohol programme that is being applied in that country.

10.2   Introduction

Anaerobic digestion of organic wastes and by-products from agriculture and the food industry is a process known for many years and is widely used for waste stabilization, pollution control, improvement of manure quality and biogas production. Anaerobic digestion is a process that exhibits many advantages: It can convert a disposal problem into a profit centre, it allows agricultural crops to be converted into a valuable fuel and it can reduce mineral fertilization demand by nutrient recovery. Therefore, anaerobic digestion has become a key method for both waste treatment and the production of renewable fuels.

During recent years, governments of many European countries as well as in other regions have increased their interest in anaerobic digestion based biogas production because it is an environmentally friendly energy source with large potential for reducing green house gas emissions. Therefore, several acts on granting priority to renewable energy sources have come into force and different governmental programs have given incentives in order to promote the development of anaerobic digestion biogas plants.

In the following, applications of biomass digestion for biogas production in the agricultural sector will be shown and discussed for the conditions in Germany and Brazil. In Europe, Germany is the leading country in this field with the highest number of installed biogas plants. In Germany, biogas is produced mainly from manure, organic waste from household, the food- and agro-industry and especially cultivated energy crops. Considerable attention will be given to Brazil. Indeed, this country has already long-term and large scale experience with the use of renewable fuel (bio-ethanol). The case of a close integration of biogas and bio-ethanol production is therefore of particular significance.

Figure 10.1 shows that the biogas yield of different substrates is strongly dependent on the type of the biomass. The fermentation of manure alone results in relatively low biogas yields. Co-digestion of manure with other wastes has a positive effect on process stability due to its high buffering capacity and its high content of trace elements. In order to increase the gas yield most of the biogas plants are operated today by co-fermentation of manure together with non-agricultural organic wastes, harvesting residues and energy crops (Figure 10.2).

Nevertheless, the treatment of organic wastes in agricultural co-fermentation plants is declining, because the regulations concerning hygiene and nutrient recycling are more stringent and the legal conditions are much more complicated as well (Weiland, 2004). Considerably higher investment and operating costs are the result, which decrease the economic benefits. The latter are coming from the entrance fee and the gas yield. On the other hand, a higher compensation is paid for the produced electricity according to the Renewable Energy Act (EEG) if only substrates from agriculture are used for

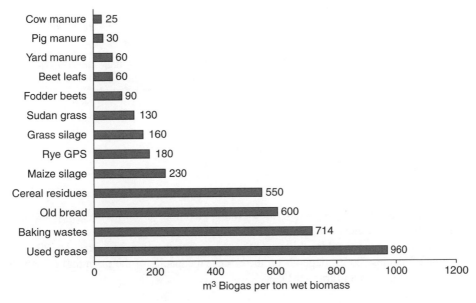

Figure 10.1 *Biogas yield from different substrates.*

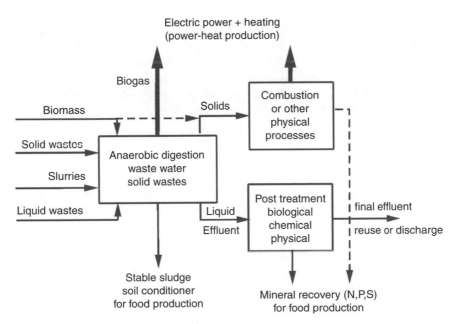

Figure 10.2 *Flow sheet of the application of anaerobic digestion for energy production and resource recovery.*

Table 10.1 The renewable energy act of the European Community provides a compensation plus a bonus which allows to obtain a total of about 15–17 €-cent per kWh electricity produced via biogas

Electrical capacity [kW]	Compensation paid for electricity [Cent/k Wh$_{el}$]	Bonus for biomass [Cent/k Wh$_{el}$]
150	11.33	6.0
150–500	9.75	6.0
500–5000	8.77	4.0

CHP bonus: 2 Cent/k Wh$_{el}$ for external heat utilization
Technology bonus: 2 Cen/k Wh$_{el}$ (e.g., dry fermentation)

biogas production (EEG, 2004). Therefore, the application of the mono-fermentation of energy crops is a fast increasing market.

The biomass bonus is paid in addition to the basic compensation for a period of 20 years. The value of biomass bonus depends on the electrical capacity of the plant and is between 4 and 6 €-Cent per kWh (Table 10.1).

10.3   Biogas Production Potential

Biogas is produced by the biological process of anaerobic digestion by which organic material is transformed into gaseous products, mainly methane and carbon dioxide. Methane production is of particular interest because it is a fuel that can be used for several applications. As shown in Figure 10.2 anaerobic digestion can be applied for biogas generation from liquid (municipal and industrial waste waters, slurries (sludges, 'liquid' manure) and solid wastes (manure and municipal refuse). A more recent liquid application is the use for biogas production from agricultural crops that are grown for the specific purpose of energy production. In the following sections, a few practical examples are presented to highlight the potential for energy generation by anaerobic digestion.

10.3.1   Germany

The Renewable Energy Act is the driving force for the development and application of agricultural biogas plants for electricity generation and the key element for climate protection, environmental protection and for the fast-growing biogas market. A study of the Federal Agricultural Research Centre (FAL) has shown that at the end of the year 2005 approx. 3000 biogas plants with a total installed electrical capacity of almost 600 MW were in operation in Germany (Figure 10.3). The total biogas potential in Germany on the basis of the available organic wastes, by-products and energy crops is calculated by the Federal Agricultural Research Centre (FAL) as 24 bil. m^3 biogas per year. The main sources are energy crops produced on approx. 1.4 mil. Hectares, manure and by-products from crop production, and processing whereas organic wastes from households and wastewater treatment have only a low potential (Figure 10.4).

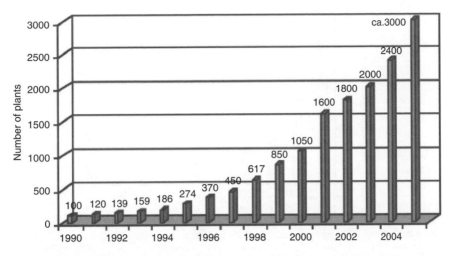

Figure 10.3 *Number of biogas plants in Germany.*

For the complete utilization of the total biomass potential, at least 20,000 biogas plants with an installed electrical capacity of approx. 300 kW are necessary.

The evaluation of modern biogas plants, which were analyzed between 2002 and 2004, showed that maize and grass silage are the most applied co-substrates in agricultural biogas plants. 80 % of all plants were operated with simultaneous fermentation of manure and maize silage and 50 % used grass silage for co-fermentation (Weiland and Rieger, 2005). More than 30 different energy-rich organic wastes from agriculture and the food- and agro-industry were applied, which were often treated simultaneously with manure and energy crops (Figure 10.5).

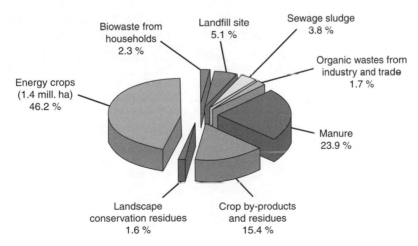

Figure 10.4 *Biogas potential and the share of substrates in Germany. (24 bill. m³ biogas per year or 13 bill. m³ methane per year, electricity 50 TWh/a and heat TWh/a, natural gas consumption 2005: 90 bill. m³.)*

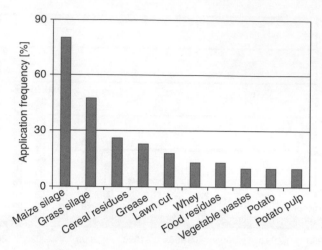

Figure 10.5 *Application frequency of various co-substrates.*

Basically all agricultural crops can be used for biogas production if the crops are not lignified and have a high yield of dry matter per hectare. Table 10.2 shows some typical crops which are suitable for biogas production.

The methane yield of the crops depends on the sort of crop, the harvesting time, the harvesting and conservation technology and many other parameters like climate and rainfall. The dry matter (DM) yield of maize is between 15 and 30 t /ha and the resulting methane yield per ton of organic dry matter (ODM) varies between 300 and 380 Nm^3/t ODM. The mean gas yields of other energy crops were published by the KTBL (KTBL, Landwirtschaftsverlag, 2005). In order to achieve high biomass yields at low input, specific crop rotation systems can be developed (Karpenstein-Machan, 2005).

10.3.2   Brazil

Although Brazil is a country with a wide variation of agricultural products and associated residues, in this paper attention will be focused on biogas production from residues of

Table 10.2   *Suitable crops for biogas production in Europe*

Annual crops	Biennial/triennial crops	Permanent crops
Maize silage	Clover-grass silage	Ley crops silage
Total cereal silage	Alfala-grass silage	
Sorghum silage	Jerusalem artichoke tops silage	
Fodder beets silage		
Sugar beets silage		
Sunflower silage		
Intercropping crops silage		
Rye corn		
Triticale corn		

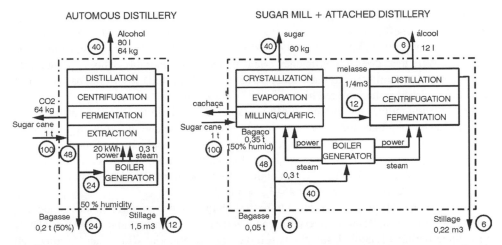

Figure 10.6   *Basic flow sheet of operations in an autonomous distillery (10.6a, left) and a sugar mill with attached distillery (10.6b. right). Encircled numbers represent the energy fraction relative to the energy in sugar cane.*

alcohol and sugar production from cane. Currently about 8 million ha are planted with sugar cane, which is used by sugar mills and alcohol distilleries in roughly equal amounts. On the average the production from cane fields is about 4000–5000 l/ha.year of alcohol or 8000 kg/ha.year of sugar (De Assis, 2006). In so called autonomous distilleries the alcohol is obtained directly from cane by juice extraction, fermentation and distillation after centrifugation to separate yeast. Figure 10.6a shows the basic flow sheet of an autonomous distillery. In sugar mills, cane juice is obtained from the cane by milling and extraction and sugar is produced by evaporation, refining and crystallization. Not all sugar in the juice is actually transformed into sugar; the last fraction of the juice is used as raw material for alcohol production at so called attached distilleries. Thus sugar mills normally are also alcohol producers. The fraction of the juice that is transformed into alcohol in these attached distilleries tends to depend on economic as well as operational factors. Figure 10.6b shows the flow sheet of a sugar mill + attached distillery.

In autonomous distilleries the yield of energy production is relatively low: only 40 % of the energy in the cane is actually converted into alcohol and whereas 48 % leaves the factory as a solid residue (bagasse) from the milling operation and 12 % is discharged from the distillation column as waste water (stillage). Current practice is to return the stillage to the cane field to use the nutrients, whereas part of the bagasse is used in steam generators to produce electric power for the distillery and the rest is sold for paper production. Bagasse is produced at a rate of 175 kgDM (almost exclusively carbohydrates) per ton of cane with a minimum humidity of 50 %. The efficiency of electric power generation at most distilleries is low due to low pressure (10 to 20 bar) of the steam generators: about half of the bagasse (1 t DM per m^3 of alcohol) is burnt to produce the required 250 kWh per m^3 of alcohol. It is known that 1 t of DM as carbohydrate roughly has an organic material mass of 1 t COD with a combustion heat of 13,9 kJ/g COD (van Haandel and Lettinga, 1994). Thus it can be calculated that the energy conversion efficiency of traditional steam generators is only 6,5 %. Distilleries using existing more modern equipment (steam pressure of 60 to

80 bar) can produce the energy with half the bagasse mass used in traditional units, so that their conversion efficiency increases to 13 %. In part these figures are so low because of the 50 % of water that makes part of the bagasse and evaporates during combustion. If the bagasses were heat dried the efficiency could increase considerably. This is particularly important at sugar mills where the energy demand is so high that most of the bagasse must be burnt to produce the required heat and power to run the mill.

In explosion motors the efficiency of biogas generators is in the range of 30 to 37 %. Thus, if anaerobic digestion of bagasse could be applied, the energy production could be increased by a factor $(30 \text{ to } 37)/13 \approx 2{,}5$ even compared to application of modern, heavy duty steam generators. If gas turbines instead of explosion motors are used for electric power generation, the efficiency is even higher (40–45 %) corresponding to an increase of energy production by a factor $(45 \text{ to } 45)/13 \approx 3 \text{ to } 3{,}5$.

Anaerobic digestion can be applied to both the liquid and the solid residues of alcohol and sugar production. It has been established that there is a release of about 500 kg of COD in stillage per m^3 of alcohol, mostly as soluble organic material which can be converted into biogas with an efficiency of more than 80 %. This means digestion of 400 kg COD and hence generation of 100 kg methane per m^3 of produced alcohol (stoichiometrically 1 kg CH_4 is produced by the digestion of 4 kg of COD). In terms of energy the methane represents about 23 % of the energy in the alcohol and 10 % of the energy in the processed cane. In conventional generators with explosion motors the electric energy that can be produced with 100 kg of methane is about 0,5 MWh.

Figure 10.7 shows the flow sheet of alcohol production with productive use of bagasse by means of combustion and of the vinasse by digestion. It is assumed that state-of-the-art equipment is used for combustion (13 % efficiency) and that efficiency of methane utilization is 35 %. The required volume of the digester can be estimated by considering a COD release of 500 kg per m^3 of produced alcohol and a loading rate of 20 kg $COD/m^3.d$, so that a volume of $500/20 = 25 \ m^3$ per $m^3.d$ of alcohol is needed for digestion during the harvest season.

It is important to stress that the digestion of the stillage increases its applicability for fertilization and irrigation: the digestion process preserves the nutrients, but the removal of the biodegradable organic matter avoids 'burning' of the sugar cane leaves. Therefore digested stillage can be applied at the most convenient moment to supply nutritional or water demand, whereas raw vinasse has to be applied just after the cane has been cut, when the cane fields are bare and there are no leaves.

Preliminary experiments show that even at environmental temperatures the digestion efficiency of bagasse is more than 50 % at a retention time of 10–12 days. The remaining 50 % is composed mainly of fibres, which could still be used for power production by combustion. Since power production is much more efficient if biogas instead of solid bagasse is used as energy source, the power output is higher if bagasse is digested. The electric power potential per m^3 of alcohol from the subproducts increases from 1,5 MWh per m^3 of alcohol if stillage digestion and bagasse combustion is used (Figure 10.7) to 2,25 MWh if bagasse is digested at 50 % efficiency (Figure 10.8).

The anaerobic digestion of bagasse in association with vinasse digestion has an important secondary advantage. Electric power can be sold at a higher price if stable production can be guaranteed throughout the year, but raw vinasse cannot be kept for a long period: after exposure of more than two months the concentration of organic material tends to decrease

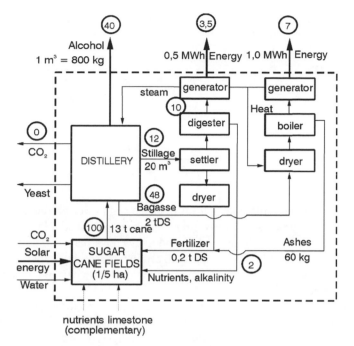

Figure 10.7 *Distillery with integral use of the sub-products by digesting vinasse and burning bagasse. Encircled numbers represent the energy fraction relative to the energy in sugar cane.*

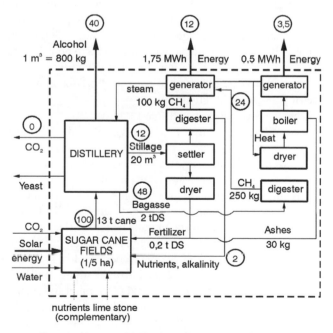

Figure 10.8 *Vinasse and bagasse digestion and burning the undigested bagasse residue. Encircled numbers represent the energy fraction relative to the energy in sugar cane.*

Table 10.3 *Demand of macronutrients for cane production required to produce of 1 m³ of alcohol and availability in the liquid and solid residues*

	kg/m3 of alcohol			% of the demand		
	N	P	K	N	P	K
Demand for production	30	5	30	–	–	–
In vinasse	4	3	20	13	60	67
In bagasse	16	0.4	1.2	53	8	4
Total in subproducts	20	3.4	21.2	67	68	71

significantly, thus reducing the potential for methane and power production. In principle, the vinasse should be processed shortly after it has been produced. However the duration of the sugar cane harvest is not more than about 180 days per year, so that during one semester per year there would be no vinasse digestion nor associated methane production. If year around energy is to be produced, one solution is to store part of the bagasse and use this material between harvests.

A retention time of 10–12 days in the reactor is required for the conversion of 50 % of the solids into methane. Consider a continuous solid anaerobic digester, for example the DRANCO type (De Baere, 2001). Assuming that the digested bagasse has a 25 % solids, this results in a loading rate of $250/12 \approx 20$ kg DM/m³.d. Because the production of 1 m³ of alcohol leads to the release of 2 t DM of biogases, the required volume is $2000/20 = 100$ m³ of solids digester per m³/d of alcohol. If power production is spread out over the whole year, the volume can be decreased to 50 m³, provided bagasse is stored: by the end of the season half of the yearly bagasse production or $0,5*4*180 = 360$ t per m³/d of alcohol (50 % humidity) must have been accumulated. It is important to note that anaerobic bagasse digestion or elutriation have not yet been implemented in continuous reactors at pilot or full scale. Therefore, at this stage the costs of bagasse digestion cannot yet be evaluated with a high degree of precision and reliability.

In Figures 10.7 and 10.8 it can be noted that the productive use of residues not only increases very significantly the output of useful energy, it also enables the recovery of a large fractions of the nutrients required for cane production. Table 10.3 shows the demand of macronutrients as well as the masses or percentages present in stillage and bagasse. Anaerobic bagasse digestion mineralizes practically all the nitrogen, so that after digestion some 67 % of the required nutrients for production are in the liquid phase and can be recovered just by recycling the effluents on the sugar cane field. Thus only about 30 % of the nutrients in a production cycle are lost and must be replenished.

Table 10.4 illustrates that the conventional alcohol production in Brazil can be increased in terms of output of products from 2775 US $ per ha per year to 3775 US $ per ha per year (factor 1.36) by incorporating the anaerobic digestion of the stillage and biogases. However, direct total digestion of the sugar cane biomass and conversion to green electricity is not yet competitive to the production alcohol for motor vehicles.

Nationwide the electric power production potential of vinasse digestion and bagasse combustion (1.5 MWh per m³ of alcohol) at the current alcohol production rate of $15*10^6$ m³/ano is 22.5 TWh/year or 2.6 GW if year around production is applied. This represents

Table 10.4 Alternative forms of energy cropping using sugar cane as the plant to generate energy. (Productivity of sugar cane = 65 $t.ha^{-1}.y^{-1}$, assumed values: alcohol: 0,5 US$/L, electric power: 0,10 US$/kWh, bagasse (50 % humidity: 10 US$/ton, nutrients: US$ 75 $ha^{-1}y^{-1}$ (67 % of total demand is considered to be recovered via bagasse and stillage)

Alternative	Method	Products (ha^{-1} $annum^{-1}$)			
		Type	Quant.	Unit	Value in US$
(A) Conventional alcohol	Alcohol fermentation + Bagasse combustion + Excess bagasse sales + Stillage recycling	Alcohol:	5000	L	2500
		El.power	1,25	MWh	125
		Bagasse	10	t (50 %)	100
		Raw stillage	100	m^3 (67 %)	50
					2775
(B) Alcohol + integral subproducts Use	Alcoholfermentation + Stillage digestion + Bagasse digestion + Combustion bagasse rest + Recycling digested stillage	Alcohol:	5000 l	L	2500
		El.power	2,5	MWh	250
		El.power	6,25	MWh	625
		El.power	2,5	MWh	250
		Digested stillage	100	m^3 (67 %)	50
					3775
(C) Total sugar cane digestion to produce electric power	Sugar cane + total anaerobic digestion + Combustion bagasse rest + Recycling digested stillage	El.power	18,75	MWh	1875
		El.power	2,5	MWh	250
		Digested effluent	100	m^3 (67 %)	50
					2175

about 4 % of the total demand of electric power in Brazil. An additional advantage of the power production from residues at alcohol distilleries is the diffuse generation, since distilleries are operating in most parts of the country. Thus the need for transmission lines is reduced.

Figures 10.7 and 10.8 show that the productive use of bagasse and stillage for energy generation transforms the use of sugar cane for alcohol production in a sustainable industry where solar energy is transformed into useful energy forms: automotive (alcohol) and electric energy with minimal use of materials for plant growth and a small environmental impact. The energy balance can be improved further by abandoning the actual practice of burning the cane before cutting to remove the leaves and the (sugar poor) tops and thus making harvesting easier. By harvesting also leaves and tops the bagasse production can be doubled, which would lead to an increase of 1,25 MWh by digestion and 0,5 kWh by combustion taking the total to 4 MWh per m^3 of alcohol, representing 28 % of the energy in the processes sugar cane.

The feasibility of productive use of the subproducts depends essentially on a comparison of investment and operational costs on the one hand and the value of the produced energy and recovered nutrients on the other. In Brazil electric power is mainly generated at hydropower plants and can be bought at prices as low as US$ 40 per MWh on the spot market, although power from the regular grit costs about US$ 80, even for large consumers. The Brazilian government is developing a special program to stimulate the production of 'green' energy (PROINFA) to complement the production of electric hydro power. Presently most of the hydroelectric power potential is already in use and additional power is being developed using fossil fuels, mostly in the form of natural gas. The expectation is that green energy will be bought by the government at US$ 80 to 100/MWh. Under those conditions there would be room for a comfortable profit for the distilleries by using the subproducts for electric power production.

In the above evaluation the benefit of CO_2 emission reduction by using renewable energy has not yet been considered. The effect can be estimated as follows: The production of 1.5 MWh of electric power per m^3 of alcohol is effected by using 2 t dry bagasse and the methane production of 100 kg of methane from vinasse digestion. If natural gas were used to produce energy (at 35 % efficiency), about 300 kg CH_4 would be required to produce 1.5 MWh and in the process 825 kg CO_2 would be released. If renewable resources are used for power production the CO_2 release is 'pre-paid' in the sense that no new CO_2 is emitted, because the organic material of the subproducts was produced by photosynthesis, using CO_2 as the carbon source. While the incentive for decreased CO_2 emission has not yet been fixed, its value is expected to be 4 to 5 US$ per ton of CO_2, or US$ 16 to 20 per ha of cane field per year in the case of autonomous distilleries and US$ 3 to 4 /ha.year for sugar mills with attached distilleries. While the absolute value of these figures is not considerable in relation to the value of the main products (sugar, alcohol or electric power), it may be significant in terms of profitability of the companies.

It is interesting to analyze the perspective of energy production if electric power generation rather than automotive fuel production is the prime objective. In that case it probably would not be advantageous to apply fermentation and the cane juice would be digested directly, whereas the bagasse would receive the same treatment as proposed above: combustion, eventually preceded by anaerobic digestion. In that case the entire equipment of the distillery would become redundant. If it is assumed that the organic material of the liquid phase (52 %) can be almost completely digested and that bagasse (48 % of the organic material) has a 50 % digestion efficiency, and knowing that 1 t of cane (with a COD of 168 kg in the liquid and 155 t in the solid phase) has a methane producing potential of $(168 + 0.5*155)/4 = 61$ kgCH$_4$, with a power generating potential of $5*61 = 307$ kWh. Combustion of the residual bagasse could yield another 77 kWh so that the total electric power production would be 384 kWh per t of cane or $65*384 = 25$ MWh/ha.year or a constant production of 2,8 kW/ha. Depending on the demand for electric power, possibilities to generate it by other means and the price that can be paid for the commodity, this may be an interesting alternative. The perspective of growing plants for electric power generation is not limited to cane: other plants such as certain types of cactus actually have a higher production potential of organic material per ha and can be grown in regions with much less favourable climate conditions and their composition (much less fibre) may allow a higher digestion efficiency.

10.4   Biogas Production Configurations

In the previous sections it was shown that the biogas production potential for renewable energy is very significant, although by itself it is insufficient to substitute the main energy sources in the matrix that is presently used. The reactor configuration to be used to unleash the potential of energy production as biogas is of great importance, since the investment in construction of the reactors represents a large fraction of the total production costs. The selection of the reactor configuration for anaerobic digestion of wastes depends first of all on its suspended solids content.

10.4.1   Configurations for Wastewater Digestion

For predominantly liquid wastes (with less than, say, 1 g/l of settleable solids) such as stillage, it is convenient to use one of the so called high rate digestion systems, which are characterized by two main design characteristics: (1) there is a mechanism for retention or recirculation of sludge, so that a high mass of active microorganisms is maintained in the reactor and (2) intense contact between influent organic material and the sludge mass is effected by appropriate measures of flow direction or mixing of the reactor contents. The first measure effectively means that the retention time of the solids or sludge age becomes independent of the liquid retention time. This means that the sludge age rather than the liquid retention time is the relevant design parameter. Hence the wastes can be treated at very short liquid retention times as long as the sludge retention mechanism is able to maintain an adequate sludge mass (compatible with the sludge age) in the reactor. The most well known high rate anaerobic digester is the UASB reactor (upflow anaerobic sludge blanket reactor, Lettinga et al., 1980), which is schematically shown in Figure 10.9.

The most characteristic device of the UASB is the phase separator, placed in the upper section and dividing the reactor in a lower part, the digestion zone, and an upper part, the settling zone. The waste water is introduced as uniformly as possible over the reactor bottom, passes through the sludge bed and enters into the settling zone via the apertures between the phase separator elements and is uniformly discharged at the surface. The biogas produced

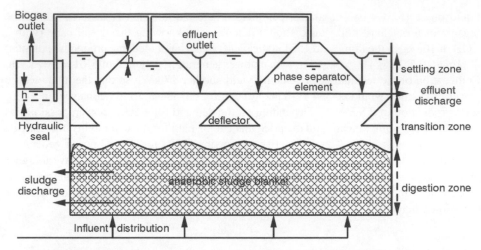

Figure 10.9 *Schematic representation of a conventional UASB reactor for anaerobic digestion of wastes with a low suspended solids concentration.*

in the digestion section is captured by the separator, so that unhindered settling can take place in the upper zone. To avoid blocking of the biogas outlet and allow separation of biogas bubble from sludge particles, a gas chamber is introduced under the separator element. The settled sludge particles on the separator elements eventually slide back into the digestion zone. Thus, the settler enables the system to maintain a large sludge mass in the reactor, while an effluent essentially free from the suspended solids is discharged. Recently several variants of the UASB concept have been developed, basically to make more efficient use of the reactor volume of the digestion section of the reactor (EGSB, Expended Granular Sludge bed reactor), and/or to increase the efficiency of the sludge retention in the settling section. Under favourable operational conditions, the UASB reactor can receive an organic volumetric load of 20 kgCOD/m^3.d. UASB reactors and its variants are currently used in a variety of liquid waste from agro business (stillage, paper mill, fruit canning effluents, tanneries, slaughterhouses) and industry (breweries, pharmaceutical, petrochemical wastes)

10.4.2   Different Process Configurations for Wet Digestion Fermenters

If the suspended solids concentration of the material to be digested is not low, biogas can be produced by wet- or dry-fermentation processes (see Figure 10.10). Today about 90 % of German agricultural biogas plants are operated by wet fermentation, which allows for a continuous treatment of liquid, pasty or solid substrates. The typical total solids content of the digester content is in a range of between 8 and 10 % TS which permits conventional pumps and mixers to be used. Dry-fermentation processes are operated with solid substrates with a resulting total solids content in the digester of at least 20 % TS. These processes are mainly operated discontinuously without any mixing of the solid substrate (Weiland, 2004b). Dry fermentation is of increasing interest because an extra technology

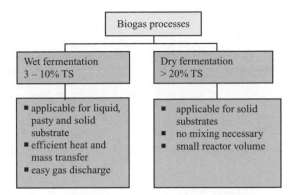

Figure 10.10 *Basic process advantages for anaerobic digestion of wastes with a high suspended solids concentration*

bonus of 2 €-Cent/kWh is paid according the Renewable Energy Act in order to stimulate this technology. The definition of dry fermentation is actually under discussion.

The most common reactor configuration employed for wet fermentation is the vertical continuously stirred tank fermenter which is applied in nearly 90 % of modern biogas plants (see Figure 10.11). Different types of mixers are applied in order to avoid scum formation or bottom layers. Central mixers with stirrer blades at different heights are very efficient if solid substrates should be mixed with manure or recycled process water. If submerged mixers are applied, at least two stirrers are necessary in order to avoid the formation of scum or bottom layers. Pneumatic mixers with recirculation of biogas are only applied in a few large ferrine litres but with negative results in the mono-fermentation of maize. The typical size of completely mixed digesters is in a range from 1000 to 4000 m³ reactor volume. Horizontal plug-flow fermenters are mainly applied in two-stage processes. For the high loaded first stage due to technical and economical aspects the reactor volume is limited to a maximum of about 700 m³.

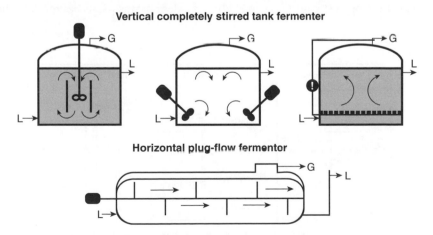

Figure 10.11 *Fermenter types for wet fermentation.*

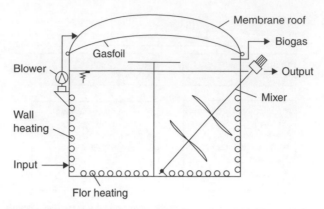

Figure 10.12   *Biogas digester with double membrane roof.*

More than 50 % of the vertically stirred tank fermenters are covered with a single or double membrane roof in order to store the gas in the top of the fermenter (see Figure 10.12). For these digester types submerged stirrers or special long axle stirrers with inclined or horizontal arrangement are applied for feeding solid wastes or energy crops into the fermenter. Special direct feeding systems on the basis of feed screws, feed pistons or flushing systems have been developed recently (see Figure 10.13).

In contrast to the conventional mixing of solid and liquid substrates in external tanks, direct feeding has the advantage that a lower energy demand is necessary for mixing, and energy losses by methane emissions can be avoided. The feeding of silage should be done with at least 12 to 24 charges per day, because the high lactic acid concentration and the low pH value can have a negative influence on process stability and gas yield (Helffrich, 2004).

Today about 2/3 of the newly erected biogas plants in Germany are designed with two fermenters in series, which result in a higher gas yield than biogas plants with only one stage. Single stage processes have the disadvantage that a part of fresh substrate can directly flow into the storage tank, which reduces the gas yield and enhances the emission of methane from open storage tanks. A study of modern biogas plants has recently shown

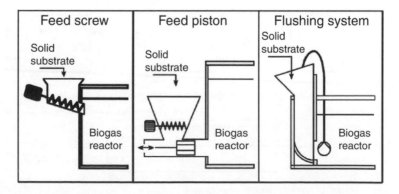

Figure 10.13   *Direct feeding systems for solid wastes and energy crops.*

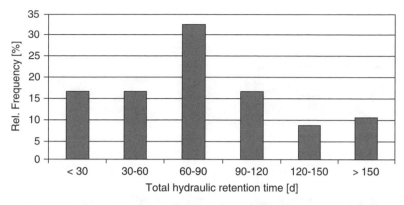

Figure 10.14   *Hydraulic retention time in modern co-digestion biogas plants.*

that the methane losses from open storage tanks are between 5 and 15 % for more than 50 % of the evaluated biogas plants (Weiland and Rieger, 2005). Only about 30 % of the biogas plants, which were built since 2004, are equipped with a gas tight storage tank. The emission of methane from the storage tank can considerably reduce the positive climate effect of biogas production, because the global warming potential of methane is a factor 21 times higher compared to carbon dioxide (Houghton, 2001). Therefore, processes with two fermenters in series equipped with a direct feeding system and a gas-tight covered storage tank should be preferred for the treatment of energy crops.

Most of the biogas plants, which use energy crops for co-fermentation, are operated with relatively low loading rates of between 1 and 3 kg $COD/m^3.d$ because fibrous material is degraded at relatively slow rates. The typical retention time of biogas plants which treat energy crops together with manure and organic wastes are between 60 and 90 days (Weiland and Rieger, 2005). Figure 10.14 shows the broad range of hydraulic retention times of modern biogas plants, The relatively long retention times result from low loadings combined with direct feeding of biomass with high total solids contents. Retention times lower than 30 days are only used for substrate mixtures of manure with a low share of energy crops.

10.4.3   Different Process Configurations for Dry Digestion Fermenters

Dry fermentation is a relatively new application of anaerobic treatment processes in the agricultural sector, which is becoming more attractive for the treatment of yard manure from cows, pigs and poultries but also for the mono-fermentation of energy crops. Several batch processes with percolation and without mechanical mixing have been tested on a pilot-scale, but only a few concepts are currently being applied on a farm-scale (Hoffmann, 2003). Figure 10.15 shows the typical process steps for dry fermentation in digestion boxes. The substrate is loaded batchwise in a closed box and mixed with inoculum from a previous batch digestion. The necessary ratio of solid inoculum has to be determined individually for each substrate. While yard manure from cows requires only small ratios of solid inoculum, up to 70 % of the input is necessary for energy crops (Kusch and Oechsner, 2005). During the digestion period, process water is recirculated and sprinkled over the

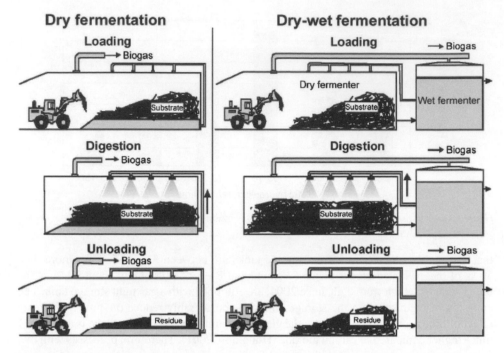

Figure 10.15 *Process variations for dry fermentation in digestion boxes.*

substrate to facilitate start-up and inoculation as well as for moisture control, heating and removal of volatile fatty acids.

To achieve a constant gas production, at least three fermenters must be operated in parallel with different strut-up times. In order to reduce the ratio of solid inoculum and to increase the process stability, the leachate should be exchanged between new and established batches, if the dry fermenter is coupled with a wet-digestion system (dry-wet fermentation), the effluent from the wet digester is used for leaching. To avoid clogging of the leach bed solids from the wet-fermenter effluent must be separated. When the digestion process is complete, the digested material is unloaded and a new batch is initiated. Another approach is to apply a continuous dry fermentation process for feedstocks that contain more than 25 % of dry matter. This is applicable to the fermentation of maize, sunflower and other types of drier energy crops, possibly mixed with high-solids containing manures.

Figure 10.16 illustrates a plant utilizing a dry fermentation technology in a continuous mode, as used for the organic fraction of municipal solid waste, but specifically adapted to energy crops. The maize silage is fed to a mixer, which intensively mixes the maize silage with a large quantity of recycled digestate. The mixture is heated with steam and is pumped to the top of the dry digester. The digestion takes place as the material moves from top to bottom due to the extraction of digestate from the bottom of the digester for either recycling in the mixer or for return to the farm land. No mixing or percolation is needed during the digestion process. For the production of 500 kW of electricity, a digester of 1200 m3 volume only is needed. The plant is operated under thermophilic conditions as dry fermentation requires less heating when operated at sufficiently high

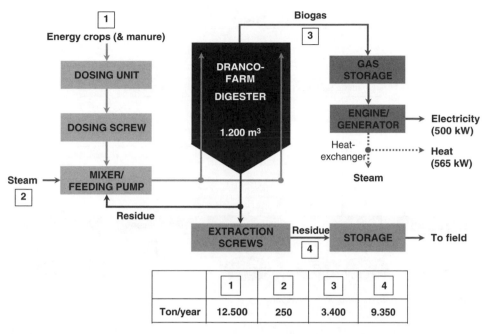

Figure 10.16 *A continuous dry fermentation process for feedstocks containing more than 25 % of dry matter. (Courtesy OWS, Belgium.)*

rates. Thermophilic digestion also yields higher degrees of degradation, resulting in an increase of 5 to 10 % of biogas production per ton.

10.4.4 Biogas Utilization

Biogas is an energy resource that can be used for many applications if the gas quality is adapted to the specific requirements (Figure 10.17). Today biogas is used in all German biogas plants in combined heat and power units (CHP) using dual-fuel injection engines or gas engines after desulphurization and removal of water. The electricity is fed into the public grid and the thermal energy is mainly used for the process and heating purposes in the house and farm but relatively seldom for decentralized public grid heating or drying processes. The efficiency of electricity generation ranges from 30 to 37 % without significant differences between the different combustion systems of the engines (Weiland and Rieger, 2005).

For the utilization of biogas as a vehicle fuel and for the feeding of biogas into the public grid or for the utilization in fuel cells much higher quality requirements have to be fulfilled. The main parameters that may require removal in upgrading systems are CO_2, H_2S, NH_3, water and solid particles. The utilization of biogas as a vehicle fuel and the feeding of bio-methane into the public gas grid have been applied in Sweden and Switzerland for several years and recently also in Germany.

A number of technologies are available for biogas upgrading. Carbon dioxide is mainly removed by water scrubbing, pressure swing adsorption (PSA) and polyethylene glycol

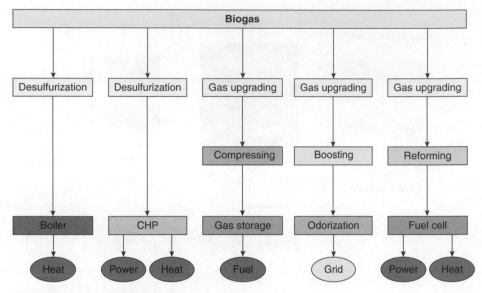

Figure 10.17   Alternatives for biogas utilization.

scrubbing (Wellinger, 1998; Weiland, 2003). H_2S can be removed internal to the digestion process by biological desulphurization performed by microorganisms of the family *Thiobacillus* or by iron chloride dosing to the digester (Schneider, 2002). Using water-scrubbing systems, H_2S can be removed simultaneously with CO_2, whereas for PSA systems adsorption columns with activated carbon are usually employed for H_2S removal. If non-agricultural wastes are employed for biogas production, higher hydrocarbons, halogenated hydrocarbons and organic silicon compounds can also be present in biogas. They can be removed by adsorption with molecular sieves or absorption in a liquid medium.

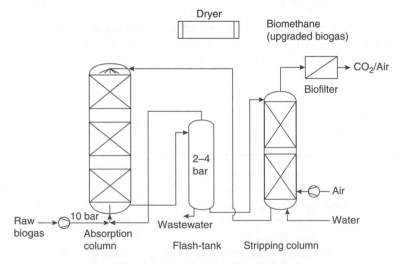

Figure 10.18   Pressure water scrubbing with regeneration of applied water.

Table 10.5 Quality specification of bio-methane as a vehicle fuel

Components	Composition
Methane	96 vol %
Carbon dioxide	< 3 vol %
Hydrogen sulphide	< 5 mg/Nm^3
Oxygen	<0,5 vol %
Particles	< 1 μm
Water content	< 32 mg/Nm^3

Water scrubbing is the most commonly applied process for carbon dioxide removal. The regeneration of the loaded water is achieved by de-pressuring and air stripping. The schematic flow sheet of water scrubbing for the removal of CO_2 and H_2S with water recirculation is presented in Figure 10.18.

There are no specific requirements defined for fuel gas, but the gas quality must be adapted to the standards of natural gas of type H. For the utilization of biogas as a vehicle fuel, the requirements specified in Table 10.5 must be fulfilled.

Upgraded biogas is a very clean vehicle fuel with respect to environment, climate and human health, because carbon dioxide, hydrocarbon and NO_x emissions are strongly reduced and no emissions of carcinogenic compounds occur. Therefore, biogas is fully free of fuel taxes in Germany and other European countries. If the upgraded gas is produced only by energy crops, the energy yield per hectare is higher by a factor of approximately four compared to rapeoil- methylester (RME), because from one hectare up to 4800 L of diesel equivalent can be obtained from biogas, but only 1200 L of diesel equivalent from RME.

The utilization of biogas in fuel cells has the advantage that the conversion efficiency to electricity is much higher compared to motor engines, but the transformation of bio-methane into hydrogen by catalytic steam reforming necessitates higher quality specifications than vehicle fuel gas. Depending on the fuel cell type, the tolerance against H_2S, NH_3, and CO is very low (Lens et al., 2005). Therefore, efficient removal of these trace gases is necessary for fuel cell applications. Due to the high investment costs and the necessarily high gas quality only a few fuel cells are applied today in practice.

10.5 Outlook

Concepts about the biological production of methane from energy crops date already from 25 years ago (Zaltman et al., 1974; Clausen et al., 1979). The current updates on two cases, Germany and Brazil, certainly demonstrate that agriculture is effectively capable to contribute considerably to the production of renewable energy. Yet, it is also well known that the agricultural route is only capable of providing a few per cent at most of the required energy supply needed a in the industrialized countries (ca. 12 L of fuel per inhabitant per day, Schnoor 2006). Producing a high fraction of the energy demand as biogas from energy crops is only feasible in countries with particularly favourable conditions like, for example, Brazil (large area, high temperature, abundant rainfall in many regions). Moreover, all countries will also continue to rely on agriculture for food and other commodities. The latter

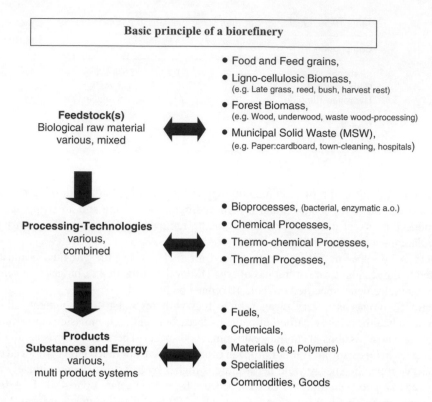

Figure 10.19 *Scheme biorefinery (after Kamm and Kamm, 2004), with kind permission of Springer Science+Business Media.*

outlets generally have higher benefit/costs ratio's. Hence, the biomass to biofuel conversion processes will be subjected to constant economic re-evaluation. A major strength of the biogas route is that it does not need to receive high-quality inputs. Indeed it can deal in a most elegant way (the biogas distils off by itself) with an almost endless variety of single and combined substrates, including materials with negative values such as wastes and residues. Hence, biogas technology will in the coming years gain in importance as a downstream process, not only in conventional agriculture and its process industries, but also in the so-called biorefineries (Verstraete et al., 2005) (Figure 10.19). In this context, it is remarkable to see that most of the concepts on biofuels do not refer to the fact that they will have wastestreams (see e.g. Hammerschlag, 2006). Yet, preliminary reports from practice indicate that both from bio-ethanol and biodiesel some 1 to 5 % of the organic matter input (COD input) will become in the biorefinery a waste for which anaerobic digestion to biogas appears to be a potential solution. These percentages generate ample prospects for the biogas technology in the downstream of the biorefinery.

A number of specific aspects can advance the position of biomass to biogas digestion in the overall industry.

First of all, there is a potential for the development of better biogas crops. For centuries, plant breeding has focussed on various aspects such as protein content, starch quality, taste, colour, aroma, disease resistance. The crop which is optimally suited for biogas production should have the following characteristics: it should mainly produce rapidly convertible carbohydrates (sugars, starch, cellulose) and contain a minimum of lignin, hemi-cellulose, proteins, minerals, pigments. Equally important is that the mass of produced organic matter per unit are is high, which depends basically on three factors: (1) the climate (2) the land and agricultural techniques applied and (3) the nature of the plant and quality of the seed. In other words, it must maximize the conversion of sunlight to digestible carbohydrate. At present, of the 15,000 kW power of sunlight which one ha of corn receives, hardly 5 is trapped as potential bioethanol power, and a mere 1.5 kW (1 out of 10 000) is recovered as effective bioethanol fuel. To the best of our knowledge, no conventional breeding programme to generate an optimal energy/biogas producing crop has ever been set up. New methods of crop 'design' implementing various genetic potentials in a specific plant used for biogas production, should be able to considerably increase the harvesting of the sun energy. At present, considerable effort is underway to design new types of food crops; it stems to reason to expect rapid advances in the design of new biogas-dedicated crops.

A second line of progress will be the development of new technologies to produce biogas. Certainly, the current techniques to harvest and preserve the biomass before it is submitted to biogas digestion need to be tuned to the subsequent and overall rate limiting methane production process. With respect to anaerobic digestion itself, two lines of reactor configuration are of value A first line relates to intensive digestion in highly technical hardware, as described above. Plenty of novel concepts will certainly arise and allow to enhance the overall energy recovery. Of special interest is for instance the coupling of anaerobic digestion with subsequent further removal of the residual organics in microbial fuel cells (Rabaey and Verstraete, 2005) Yet, this approach will demand major investments. Although on average a digester constitutes only a relative simple amount of infrastructure representing on average an all-in capex of 1000 Euro per m^3, its payback at the current energy price of methane (0.2 Euro per m^3) and reactor rate of methane production (10 volumes of methane per volume of reactor per day) is still of the order of several years. Hence, it stems to reason to also consider the possibility to harvesting biomass, bringing it together in large scale landfill type reactor system and then utilize this biomass, stored in a less expensive way, over a longer time scale (Verstraete et al., 2005).

A third line of improvement must be sought in the better positioning of the biogas produced. At present, the biogas is generally combusted to render electricity. Technology to upgrade it to vehicle fuel is already long time available (Henrich, 1981). Yet, provided it is produced at sufficiently large scale, a thermal conversion to syngas (H_2 and CO) is quite possible (Effendi et al., 2005; Wang and Chang, 2005). The advantage of this route is that the conventional petrochemistry can directly tap on this resource and produce from it all possible commodities. Once biogas digestion is linked to conventional petrochemistry, it will have a major inroad to the chemical industry and hence the overall industrial society.

10.6 Conclusions

Biogas from agriculture has several strong points. It can deal with a variety of materials and it thus fits perfectly in the various routes of downstream processing of the agro-business. Secondly, it allows recovering energy with a maximum of efficiency because its end product, methane, distils by itself from the mixture which is used to produce the fuel. Third, it can be used in a number of ways, going from direct combustion, to conversion to hydrogen and electricity and to cracking and becoming syngas to produce commodity chemicals and materials. Fourth, the technology can be designed at small, medium and even super-large scale in case of landfill storage projects. Finally, and certainly also of critical value, the anaerobic digestion of biomass to biogas allows to recover all plant nutrients and surpasses all other biomass conversion technologies in terms of environmental sustainability, because it can harvest the sun and at the same time return to the soil the undigested plant residues, in the lignin, which will improve the soil organic matter content and the minerals which will allow to grow the next crop without input of major amounts of external fertilizer.

References

Clausen, E.C., Sitton, O.C. and Gaddy, J.L. 1979. Biological production of methane from energy crops. *Biot Bioeng* **21**: 1209–1219

De Assis, P.E.P. 2006. *Integrated production of sugar and alcohol. Int. Workshop on production and use of alcohol*. Havana, Cuba.

De Baere, L. 2001. Full-scale experience with digestion facilities treating 25.000 and 50.000 ton of biowaste per year. In : Proceedings (Part 2) of 9th World Congress Anaerobic Digestion: 'Anaerobic Conversion for Sustainability', Antwerp.

EEG, Erneuerbare-Energien-Gesetz. 2004. Bundesgesetzblatt, Teil I, Nr. 40. Bonn.

Effendi, A., Hellgardt K., Zhang Z.G. and Yoshida, T. 2005. Optimising H_2-production from model biogas to combined steam reforming and CO shift reactions. Fuel **84**, 869–874.

Hammerschlag, R. 2006. Ethanol's energy return on investment : A survey of the literature 1990 – present. Environ. *Sci; Technol.* 1744–1750.

Helffrich, D. 2004. Lagerung, Einbringung und Rühren nachwachsender Rohstoffe zur Vergärung in landwirtschaftlichen Biogasanlagen: Biogas – zuverlässige Energie von Wiese und Acker, Fachverband Biogas, *Freising*, p. 64–67.

Henrich, R.A. 1981. Municipal waste to vehicle fuel. *Biocycle* May-June, 27–29.

Hoffmann, M. 2003. Trockenfermentation in der Landwirtschaft: Entwicklung und Stand, in VDI-Berichte 1751, *VDI-Verlag, Düsseldorf*, p. 193–201.

Houghton, J.T. 2001. *Climate Change : The Scientific Basis*, Cambridge University Press, Cambridge.

Kamm, B. and Kamm, M. 2004. Principles of biorefineries. *Appl. Microbiol. Biotechnol.* **64**, 137–145.

Karpenstein-Machan, M. 2005. Energiepflanzenbau für Biogasanlagenbetreiber, *DLG-Verlag*, Frankfurt.

Kuratorium für Technik und Bauwesen in der Landwirtschaft (KTBL), *Gasausbeute in landwirtschaftlichen Biogasanlagen, Landwirtschaftsverlag*. 2005. Münster.

Kusch, S. and Oechsner, H. 2005. *Biogas production in discontinuously operated solid-phase digestion systems*. Proc. of the 7th FAO/SREN-Workshop, Uppsala, Sweden.

Lens, P., Westermann, P., Haberbauer, M. and Moreno, A. 2005. *Biofuels for Fuel Cells*, IWA Publishing, London.

Lettinga, G., van Velsen, A.F.M., Hobma, S.W., de Zeeuw, W. and Klapwijk, A. 1980. Use of the upflow sludge blanket (USB) reactor concept for biological wastewater treatment, Especially for anaerobic treatment. *Biotechnol. Bioeng.* **22**, 699–734.

Rabaey, K. and Verstraete, W. 2005. Microbial fuel cells: novel biotechnology for energy generation. *Trends in biotechnology* **23**, 291– 298.

Schneider, R. 2002. Grundlegende Untersuchungen zur effektiven, kostengünstigen Entfernung von Schwefelwasserstoff aus Biogas – Biogasanlagen: Anforderungen zur Luftreinhaltung, *Bayerisches Landesamt für Umweltschutz*, Augsburg, p. 25–41.

Schnoor, J.L. 2006. Biofuels and the environment. *Env. Sci. Technol.* **1**: 4042.

van Haandel, A.C. and Lettinga, G. 1994. *Anaerobic sewage treatment in regions with a hot climate.* John Wiley & Sons, Chichester, United Kingdom (1994).

Verstraete, W., Morgan F., Aiyuk, S., Waweru, M., Rabaey, K. and Lissens, G. 2005. Anaerobic digestion as a core technology in sustainable management of organic matter. *Wat. Sci. Technol.* **52**: 59–66.

Wang, T.J. and Chang, L.P. 2005. Synthesis gas production via biomass catalytic gasification with addition of biogas. *Energy & Fuels* **19**, 637–644.

Weiland, P. 2003. Notwendigkeit der Biogasaufbereitung, Ansprüche einzelner Nutzungsrouten und Stand der Technik, in *Gülzower Fachgespräche*, Bd. 21, Gülzow, p. 7–22.

Weiland, P. 2004a. Ist die gemeinsame Vergärung von Bioabfällen, landwirtschaftlichen Abfällen und Energiepflanzen möglich und sinnvoll? *EEG und Emissionshandel*, ORBIT, Weimar, p. 89–100.

Weiland, P. 2004b. Stand der Technik bei der Trockenfermentation: Zukunftsperspektiven, Gülzower Fachgespräche, Bd. 23, Gülzow, p. 23–35.

Weiland, P. and Rieger, C. 2005. Ergebnisse des Biogas-Messprogramms, Fachagentur Nachwachsende Rohstoffe, Gülzow.

Wellinger, A. 1998. Biogas upgrading and utilization, in IEA Bioenergy, Task 24: Energy from Biological Conversion of Organic Material, Sailer Druck, Winterthur.

Zaltman, R., Doner, D. and Bailie, R.C. 1974. Perpetual methane recovery system. *Compost Sci.* **15**: 14–19.

11

Biological Hydrogen Production by Anaerobic Microorganisms

Servé W.M. Kengen, Heleen P. Goorissen, Marcel Verhaart and Alfons J.M. Stams

Laboratory of Microbiology, Wageningen University and Research Center

Ed W.J. van Niel

Laboratory of Applied Microbiology, University of Lund, Sweden

Pieternel A.M. Claassen

Agrotechnology and Food Sciences Group, Wageningen University and Research Center, Wageningen, The Netherlands

Abbreviations

Ca	*Caldicellulosiruptor*
Cl	*Clostridium*
E.	*Escherichia*
Eb	*Enterobacter*
ED	Entner-Doudoroff
EM	Embden-Meyerhof
P.	*Pyrococcus*
$P(H_2)$	Hydrogen partial pressure
PP	Pentose phosphate
Ta	*Thermoanaerobacter*
Tt	*Thermotoga*

Biofuels Edited by Wim Soetaert and Erick J. Vandamme
© 2009 John Wiley & Sons, Ltd

11.1   Introduction

Hydrogen gas has great potential as a future fuel, as during its combustion the greenhouse gas CO_2 is not produced. However, hydrogen is only a true 'green' fuel if it is produced from renewable sources like wind, sunlight, geothermal energy or biomass. A wide variety of microorganisms is able to form hydrogen in light-dependent and in light-independent processes, such as dark fermentations. Hydrogen can be formed in the fermentation of complex biomass, but for large-scale production thus far only hydrogen formation from polysaccharides is feasible. The various types of microorganisms that can play a role in hydrogen formation by dark fermentations will be discussed here, with special emphasis on the thermophilic hydrogen-producing microorganisms. At elevated temperatures hydrogen formation is thermodynamically more feasible, and less undesired side-products are produced. The current knowledge on the key enzymes (hydrogenases, oxidoreductases) of the various hydrogen producing microorganisms is highlighted, making use of the genome information that is available now for several of the hydrogen producing species. The availability of the complete genome sequences also offers the possibility to apply genetic tools to optimize hydrogen formation. Acetate is an obligate end product of dark fermentations, which limits the hydrogen yield. Options to deal with the acetate problem are discussed. The application of electricity-mediated electrolysis would broaden the range of compounds that can be used for hydrogen formation.

11.2   Hydrogen Formation in Natural Ecosystems

In methanogenic environments hydrogen is a key intermediate in the anaerobic decomposition of organic matter to methane and carbon dioxide.[1] At low temperatures (up to about 45 °C) about 1/3 of the methane is formed by reduction of carbon dioxide with hydrogen as electron donor:

$$4\,H_2 + HCO_3^- + H^+ \rightarrow CH_4 + 2\,H_2O \qquad \Delta G^{0\prime} = -135.6\,\text{kJ/mol methane}$$

while 2/3 of the methane is formed by acetate cleavage:

$$(\text{acetate}^- + H_2O \rightarrow HCO_3^- + CH_4 \qquad \Delta G^{0\prime} = -31\,\text{kJ/mol methane}$$

As methanogens cannot metabolize complex organic carbon compounds (polysaccharides, proteins, lipids, nucleic acids, etc), fermentative bacteria are required to funnel the degradation of complex organic compounds to the methanogenic substrates hydrogen and acetate. Irrespective the type of microorganisms that are involved and irrespective the nature of organic compounds, in methanogenic environments methane and carbon dioxide are the final carbon end products. At moderately high temperatures (50–80 °C), even all methane is formed from the reduction of carbon dioxide by hydrogen. This is because at these conditions, acetate is first degraded by bacteria to form hydrogen and carbon dioxide:

$$\text{acetate}^- \rightarrow 2\,HCO_3^- + H^+ + 4\,H_2 \qquad \Delta G^{0\prime} = +104.6\,\text{kJ/mol acetate}$$

the hydrogen being used by methanogens to reduce carbon dioxide to methane.[2]

   At first glance it may seem that at high temperatures all organic carbon can be converted into carbon dioxide and hydrogen, provided that methanogens are inhibited. This indeed is the case but hydrogen is only formed when it is efficiently taken away by methanogens. Thermodynamically, acetate can only be oxidized to carbon dioxide at a very low hydrogen partial pressure $(P(H_2))$. Thus, hydrogen inhibits its own formation. This is not only the case for acetate, but also for a variety of other organic substrates.[1]

Thus far, formation of hydrogen-rich gas in dark fermentation is only feasible from sugars, although it is restricted by thermodynamics. This is also true for thermophilic microorganisms, even though they harbor the biochemical potential to convert complex organic carbon completely to carbon dioxide and hydrogen. Thus, energy input is needed to overcome the thermodynamic barrier.

11.3   Thermodynamics of Hydrogen Formation

Carbohydrates are the main substrates for fermentative bacteria that produce hydrogen.[3–5] Cellulose and hemicellulose are the most abundant polysaccharides available in nature, and therefore, glucose and xylose are the predominant monomeric sugars used for hydrogen formation. Starch and sucrose may also be available in large amounts as these are used as storage material in various plants. Fermentation of glucose and xylose results in hydrogen formation, but in addition, depending on the microorganism, also acetate, butyrate, lactate, ethanol and some other organic compounds (e.g. formate, butanediol, succinate) are produced. Under standard conditions complete oxidation of glucose to CO_2 and H_2 is thermodynamically not possible:

$$\text{glucose} + 12\,H_2O \rightarrow 6\,HCO_3{}^- + 6\,H^+ + 12\,H_2 \; \Delta G^{0'} = +3.2\,\text{kJ/mol}$$

The oxidation of glucose proceeds only to acetate, and even in this case other end products are usually produced as well. Intermediates of the metabolism are used as electron acceptors resulting in branched pathways, leading to butyrate, lactate, ethanol, alanine etc. (Figure 11.1). Production of acetate is coupled to the synthesis of ATP by substrate-level-phosphorylation, whereas production of the other products yields less or no ATP. These branched pathways enable the microorganism to adjust the metabolism so that an optimal ATP gain and thermodynamic efficiency of ATP synthesis are accomplished.[6]

The oxidation of glucose to 2 acetate and 4 H_2 is thermodynamically feasible, but in natural environments this only occurs when the hydrogen concentration is kept low by hydrogen-consuming methanogens.[1]

$$\text{glucose} + 2\,H_2O \rightarrow 2\,\text{acetate}^- + 2\,HCO_3{}^- + 4\,H_2 \qquad \Delta G^{0'} = -206.1\,\text{kJ/mol}$$

The need for a low hydrogen partial pressure can be explained best when we look at the hydrogen producing reactions. During catabolism (e.g. the Embden Meyerhof pathway) reducing equivalents are produced in the form of NADH (glyceraldehyde-3-P dehydrogenase reaction) and in the form of reduced ferredoxin (pyruvate:ferredoxin oxidoreductase reaction) (Figure 11.1). The midpoint redox potential of the couples $NAD^+/NADH$ and oxidized ferredoxin/reduced ferredoxin are $-320\,mV$ and $-398\,mV$, respectively.[6] These electron carriers must be recycled continuously for catabolism to proceed. Various fermentation reactions exist that can accomplish this recycling of electron mediators, but the

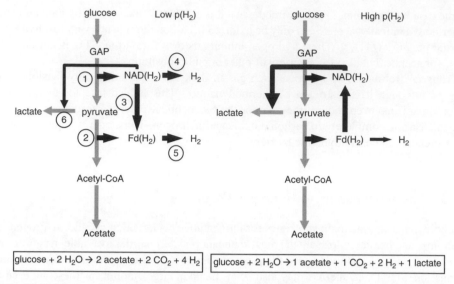

Figure 11.1   *Scheme of a typical carbon (grey)- and electron (black) flow during glucose fermentation. For simplicity pathways to ethanol, butyrate and alanine are not shown. At low P(H$_2$) reducing equivalents are transferred mainly to H$_2$, involving glyceraldehyde-3-P dehydrogenase (1), pyruvate:ferredoxin oxidoreductase (2), NADH:ferredoxin oxidoreductase (3), NADH-dependent hydrogenase (4) and ferredoxin-dependent hydrogenase (5). At high P(H$_2$) reducing equivalents are partly transferred to lactate, involving lactate dehydrogenase (6)*

conversion that is most important is determined by the Gibbs free energy change at standard conditions ($\Delta G^{\circ\prime}$) of the individual conversions (Table 11.1).

From these reactions it can be seen that the formation of hydrogen by reduction of protons with NADH as electron donor is thermodynamically unfavorable: the midpoint redox potential of the couple H$^+$/H$_2$ being −414 mV. When NADH is reoxidized only via the formation of ethanol, lactate or alanine etc., the amount of hydrogen that is produced would never exceed the amount of CO$_2$. Nevertheless, available fermentation data show that the H$_2$/CO$_2$ ratio can easily reach values higher than 1 and even reach 2, as was reported for *Thermotoga maritima* (Table 11.2).

This indicates that NADH is not only used for the exergonic dehydrogenase catalyzed reactions, but that under certain conditions NADH oxidation also results in proton

Table 11.1   *Gibbs free energy values for different fermentative reactions. Data were calculated using Thauer et al.[6], Amend and Shock[7] and Amend and Plyasunov[8]*

Fermentative reaction	$\Delta G^{0\prime}$ kJ/reaction
NADH + H$^+$ + pyruvate$^-$ → NAD$^+$ + lactate$^-$	−25.0
2NADH + 2H$^+$ + acetyl-CoA → 2NAD$^+$ + ethanol + CoA	−27.5
NADH + H$^+$ + pyruvate$^-$ + NH$_4$$^+$→ NAD$^+$ + alanine + H$_2$O	−36.7
NADH + H$^+$→ NAD$^+$ + H$_2$	+18.1
2 Ferredoxin(red) + 2H$^+$→ 2 Ferredoxin(ox) + H$_2$	+3.1

Table 11.2 Overview of anaerobic and facultative anaerobic hydrogen producing microorganisms (modified from de Vrije and Claassen, 2003 (9)

Organism	Domain	T-optimum (°C)	Culturing type	Substrate	H_2/glucose	Condition	Reference
Mesophiles							
Enterobacter aerogenes E.82005	B	38	batch	glucose	1.0		(10)
Enterobacter aerogenes E.82005	B	38	chemostat	molasses	0.7		(11)
Enterobacter aerogenes HU-101	B	37	chemostat	glucose	0.6		(12)
Klebsiella oxytoca HP1	B	38	chemostat	glucose	1.0		(13)
Clostridium butyricum	B	30	chemostat	glucose	1.21–1.64	Butyrate formed	(14)
Clostridium acetobutylicum	B	34	controlled batch	glucose	1.4		(15)
Clostridium sp. No 2	B	36	batch	glucose	2.0		(16)
Ruminococcus albus	B	37	chemostat	cellulose	1.4		(17)
		37	chemostat	cellulose	1.4		(18)
Thermophiles							
Clostridium thermosaccharolyticum	B	55	batch	glucose	0.69	pH 7	(19)
					1.63	pH 6	
			chemostat	glucose	1.44	pH 5.5	
Clostridium thermocellum JW20	B	60	batch	Cellulose	0.68	No stirring	(20)
					1.23	Stirred	
			batch	cellobiose	0.61	No stirring	
					1.01	Stirred	
			batch	glucose	1.8		
Clostridium thermocellum JW20	B	60	chemostat	glucose	3.5		(21)
Clostridium thermocellum LQRI	B	60	batch	cellobiose	1.33	No stirring	(22)
					2.86	Stirred	
Clostridium thermocellum 27405	B	60	batch	cellobiose	0.82–1.55	No stirring	(23)

(Continued)

Table 11.2 Continued

Organism	Domain	T-optimum (°C)	Culturing type	Substrate	H$_2$/glucose	Condition	Reference
Extreme/hyperthermophiles							
Thermotoga elfii	B	65	controlled batch	glucose	3.3	Stirred	(5)
		65	Batch	glucose	3.3		(24)
Thermotoga neapolitana	B	70	Batch	glucose	4	Stirred + oxygen	(25)
Thermotoga maritima	B	80	Batch	glucose	4		(26)
Caldicellulosiruptor saccharolyticus	B	70	controlled batch	sucrose	3.3	Stirred	(5)
					2.65		(4)
Thermoanaerobacter tengcongensis	B	75	batch	glucose	0.3 (2)	2 H$_2$ based on [acetate].	(27)
		75	fermentor	glucose	4	Stirred, N$_2$ flushed	(28)
Pyrococcus furiosus	A	90	batch	maltose	2.6–3.5	Stirred, alanine produced	(29)
		90	batch	cellobiose	2.4–2.8	Stirred, alanine produced	
			controlled batch	starch			(30)
			chemostat	maltose	2.9		(31)
Thermococcus kodakaraensis	A	85	chemostat	starch	3.33		(32)

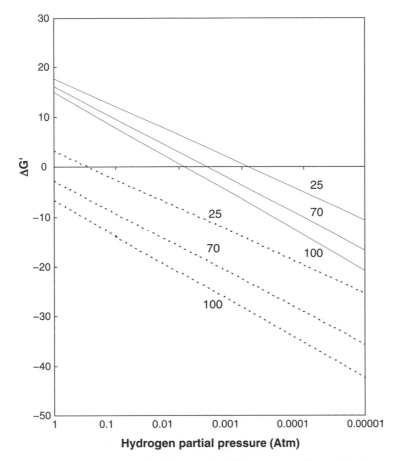

Figure 11.2   *Effect of the hydrogen partial pressure on the Gibbs free energy change of hydrogen production from NADH (solid line) and reduced ferredoxin (dashed line) at three different temperatures (25, 70 and 100 °C). Values calculated from data in (6, 34, 35).*

reduction to form H_2. There are three reasons that can explain these observations. Firstly, $\Delta G^{\circ\prime}$ values are calculated for standard conditions (1 Molar concentration of the reactants, 25 °C, pH 7), which may differ considerably from the average physiological conditions. For instance, when the $P(H_2)$ is low, the situation becomes notably different. For example, at a $P(H_2)$ of 10^{-4} atm proton reduction by NADH is even exergonic (-4.7 kJ/mol). Depending on the $NAD^+/NADH$ ratio this value may become even more negative. Secondly, also temperature affects the thermodynamics of a reaction (($\Delta G^0 = \Delta H - T\Delta S^0$)). As discussed by Stams,[33] hydrogen producing reactions become energetically more favorable at higher temperatures. Figure 11.2 illustrates how the $\Delta G^\prime$ value of proton reduction by NADH or ferredoxin changes as function of the $P(H_2)$ and temperature.

These data suggest that at elevated temperatures H_2 production from NADH is even more exergonic, which corresponds with the high H_2 levels reported for *Tt. maritima* and other hyperthermophiles (Table 11.2). A third explanation for hydrogen formation from

NADH is that the reaction might be driven by reversed electron transport. This would require that the NAD-dependent hydrogenase resides in the membrane. Such energy-driven uphill oxidation of NADH has been described for *Clostridium tetanomorphum*, which contains an NADH-dependent oxidoreductase that uses a sodium gradient to accomplish the endergonic reduction of ferredoxin. Reduced ferredoxin is subsequently used to produce H_2.[36] However, an NADH-dependent hydrogenase has been purified from *Thermoanaerobacter tengcongensis*, but this Fe-only hydrogenase is not membrane-bound.[28] Moreover, *Pyrococcus furiosus* contains two NADPH-dependent hydrogenases, which are also not membrane-bound.[37] The latter data suggest that hydrogen formation from NADH is not necessarily dependent on an energy input via reversed electron transport, or at least not in extreme- or hyperthermophiles.

Figure 11.2 illustrates that hydrogen formation from ferredoxin is thermodynamically much more favorable, especially at elevated temperatures. At $70\,°C$ hydrogen formation from ferredoxin is even exergonic in the presence of 1 atm of hydrogen.

11.4   Enzymology

Molecular hydrogen is produced by hydrogenases (EC 1.12.99.6, EC 1.12.7.2), which catalyze the reduction of protons and thereby release the reducing equivalents that are formed during the anaerobic degradation of organic substrates. Hydrogenases can also perform the reverse reaction, which allows microorganisms to use H_2 as a source of reductant. Two main types of hydrogenases can be distinguished based on their metal content, viz. the Fe-only hydrogenases and the [NiFe] hydrogenases. For details on the structural composition and the catalytic mechanism of the various hydrogenases we refer to some excellent reviews on these topics.[38–41]

Fe-only hydrogenases seem to be restricted to strictly anaerobes, whereas [NiFe] hydrogenases are found more wide-spread in anaerobes, facultative anaerobes, and aerobes. Both types of hydrogenase play a key role in the fermentative production of H_2 which is discussed here. As described above different electron carriers can deliver the electrons to the terminal hydrogenase, viz. ferredoxin, NAD(H) or NADP(H), and these electron carriers are reduced in a limited number of oxidation steps in the central metabolic pathways. The two main oxidation steps during anaerobic sugar degradation (either EM- or PP-pathway) are the conversion of glyceraldehyde-3-P to 3-P-glycerate and the conversion of pyruvate to acetyl-CoA. Recycling of the reduced carriers occurs by different enzymatic steps.

Generally, reduced ferredoxin is used for proton reduction. As shown above, this is an energetically favorable reaction which can be catalyzed by cytoplasmic monomeric Fe-only hydrogenases (e.g. *Cl. pasteurianum*)[42] or by multisubunit membrane-bound [NiFe] hydrogenases.[41] Some of these [NiFe] hydrogenases are able to convert energy and their six-subunit basic structure shows resemblance to the catalytic core of the NADH:quinone oxidoreductase (Complex I). Several of the hydrogen-producing microorganisms discussed here harbor such an energy-conserving [NiFe] hydrogenase (Table 11.3).

Hydrogen formation from NADH requires an NADH-dependent hydrogenase, which has recently been characterized from *Ta. tengcongensis* (for details see also paragraph on *Thermoanaerobacter*).[28] Homologs of the gene encoding this Fe-only hydrogenase can be identified in *Thermoanaerobacter ethanolicus*, *Caldicellulosiruptor saccharolyticus*,

Table 11.3 *Overview of oxidoreductases and hydrogenases and encoding gene clusters in some key fermentative hydrogen-producing micro-organisms. Numbers refer to locus tags. Gene clusters are given in brackets. Experimentally confirmed gene products are underlined. Fd = ferredoxin*

		Escherichia coli K12	Clostridium thermocellum	Caldicellulosiruptor saccharolyticus	Thermoanaerobacter tengcongensis	Thermotoga maritima	Pyrococcus furiosus
Membrane bound	Fd-dependent NiFe membrane-bound hydrogenase complex (Mbh, Mbx)					Mbx: 1205 → 1217	Mbh: 1423 →1436, Mbx: 1441→1453
	Fd-dependent NiFe membrane-bound hydrogenase complex (EchA-F)		Ech: 3020 → 3024	Ech: 1534 → 1539	0123 → 0128		
	NADH: Fd oxidoreductase (Nfo,Rnf)		Rnf: 2430 → 2435			Rnf: 0244– –0249	
	Formate hydrogen lyase complex (HycA → 1)	Hyc: 2717 → 2725					
Cytoplasmic	NADH-dependent Fe-only hydrogenase (HydA → C/D)		0338, 0340, 0341, 0342	1860, 1862, 1863, 1864	0890, 0892, 0893, 0894	1424, 1425, 1426	
	NADP-dependent (Sulf)hydrogenase I and II (NiFe)						H-I: 0891 → 0894, H-II: 1329 →1332
	Fd: NADP oxidoreductase (Sulfide dehydrogenase)						SuDH: 1327,1328

Clostridium thermocellum and *Tt. maritima* suggesting that also these thermophiles contain a NADH-dependent hydrogenase (Table 11.3). Alternatively, NADH may transfer its electrons first to ferredoxin by an NADH:ferredoxin reductase, and the reduced ferredoxin is subsequently used for proton reduction. An NADH:ferredoxin reductase (EC 1.18.1.3) was described by Jungermann et al.[43] and Gottschalk et al.;[44] and this activity has since been demonstrated in many anaerobic fermentative bacteria including *Tt. maritima*[26] and *Clostridium cellulolyticum*.[45] *Cl. butyricum* appears to have one bidirectional enzyme, whereas *Clostridium acetobutylicum* probably has two enzymes (forward and reverse).[46] However, despite the extensive research on these enzymes in the Sixties and Seventies, these enzymes have never been purified and gene sequences are also not known (according to BRENDA database). More recently, NADH:ferredoxin oxidoreductases have been purified from *Cl. tetanomorphum*, *Acidaminococcus fermentans* and *Fusobacterium nucleatu* and they appeared to be membrane-bound enzymes.[36] They either use the energy difference between ferredoxin and NAD to generate a Na^+ gradient, or they catalyze the reverse reaction and use a Na^+ gradient to reduce ferredoxin by NADH. Obviously, not all NADH:ferredoxin oxidoreductase interconversions occur at the membrane, because e.g. *P. furiosus* harbors two cytoplasmic ferredoxin:NADPH oxidoreductases (sulfhydrogenase and sulfide dehydrogenase)[37,47] (Table 11.3).

Hydrogen can also be produced in (facultative) anaerobes without ferredoxin or separate hydrogenases. For instance, the facultative anaerobic enterobacteria differ in their hydrogen enzymology by the sequential use of a pyruvate formate lyase and a formate hydrogen lyase.

11.5   Enterobacteria

Species of the genus Enterobacteriaceae are facultative anaerobic. In the absence of oxygen and inorganic electron acceptors they perform a mixed acid fermentation, resulting in the formation of hydrogen and other products like formate, acetate, ethanol, lactate, succinate and butane-diol. The ratio at which these products are formed is dependent on the species and the culture conditions. *Escherichia coli* produces little butane-diol, while this is a main end product of *Enterobacter aerogenes* (Table 11.4). Interestingly, *E. coli* forms formate as main product when grown at high pH, while hydrogen is only formed at a low pH. In the aerobic metabolism enterobacteria convert pyruvate to acetyl-CoA and CO_2 via an NAD-dependent pyruvate dehydrogenase complex. NADH is oxidized to NAD in the electron transport chain. However, under fermentative conditions a pyruvate formate lyase is involved in pyruvate conversion which results in the formation of acetyl-CoA and formate. Formate is converted to hydrogen and carbon dioxide by means of a formate hydrogen lyase.[49] The role of formate in hydrogen production in *E. coli* has been discussed by Sawers.[50] Formate is excreted by this bacterium, but at an acid pH it is taken up and split into hydrogen and carbon dioxide. Cleavage of formate results in energy conservation by means of a proton gradient.[51] A proton pumping hydrogenase (Ech) as described by Hedderich and Forzi[41] may be involved. Thermodynamically hydrogen formation from formate (formate$^-$ + H_2O → H_2 + HCO_3^-; $\Delta G^{0\prime}$ = +1.3 kJ/mol formate) is intriguing. The $\Delta G^{0\prime}$ of this conversion is + 1.3 kJ, which implies that this conversion is affected by the $P(H_2)$ as well.

Table 11.4   *Product formation by Escherichia coli and Enterobacter aerogenes. Products are expressed in mmoles per 100 mmoles glucose fermented (48)*

	E. coli pH6.2	E. coli pH 7.8	E. aerogenes
Formate	2.4	86.0	68.4
Hydrogen	75.0	0.3	n.d.
Carbon dioxide	88.0	1.8	79.6
Acetate	36.5	38.7	51.9
Lactate	79.5	70.0	10.1
Ethanol	59.8	50.5	51.5
Succinate	10.7	14.8	13.1
2,3 butanediol	0.3	0.3	19.2

n.d., not determined

Besides hydrogen formation from formate, NADH which is formed in the conversion of glyceraldehyde-3-phophate to 3 phosphoglycerate needs to be directed to hydrogen formation as well. In general, mass balances show that maximally only two molecules of hydrogen are formed per molecule of glucose.[52] However, mutant strains of *Eb. cloacae* in which some pathways to alcohols had been blocked, showed hydrogen yields of up to 3.4 mol per mol of glucose.[53,54] These data indicate that in the enterobacteria NADH can also be used for hydrogen formation.

Several studies have been performed to enhance hydrogen formation in enterobacteria. By redirecting biochemical pathways by means of inhibitors and by creation of specific mutants an enhanced hydrogen production was observed in *Enterobacter aerogenes*.[53,55,56] The hydrogen production yields and the hydrogen production rates were enhanced by genetically modifying *E. coli* strains.[57,58] These authors interrupted genes involved in lactate and succinate formation and overexpressed formate hydrogen lyase.

11.6   The Genus *Clostridium*

Clostridia are well-known hydrogen producers. These strictly anaerobic sporeforming bacteria are found in environments that are rich in decaying plant materials and therefore have the enzymatic machinery to hydrolyze polymers like cellulose, xylan, pectin, chitin and starch. Their high hydrolytic capacity is one of the reasons that clostridia have been investigated repeatedly for biofuel production.[59,60] In addition to hydrogen also ethanol and butanol can be produced. The optimum growth temperature ranges from 25 to 65 °C, but most are typical mesophiles with optima between 30–40 °C. A few selected species have been investigated in more detail because of the high level production of specific end products. *Cl. acetobutylicum* and *Cl. beyerinckii* perform the so-called ABE fermentation, producing acetone, butanol and ethanol.[61] *Cl. thermocellum* has been studied predominantly for its high ethanol producing capacity combined with a high cellulose hydrolyzing capacity.[62] Among the clostridia, *Cl. thermocellum* represents the one with the highest growth temperature. A draft genome sequence of *Cl. thermocellum* is available (2005) and first micro-array experiments are underway. H_2 metabolism has been studied in particular in *Cl. pasteurianum*[42,63,64] and the first hydrogenase was purified from this organism.[64] In fact all clostridial species are able to produce hydrogen to some extent, but the amount of hydrogen produced is dependent on the species and also on the growth condition.

In general, under conditions of low hydrogen pressure, carbon source limitation (low flux through glycolysis) and low growth rates, high levels of hydrogen can be produced, approaching the maximal ratio of 4 H_2 per glucose. However, in standard batch incubations these conditions are never met and substantial production of more reduced end products occurs (i.e. lactate, ethanol, butanol, acetone) instead of ATP-generating production of acids (acetate, butyrate). Table 11.2 summarizes the hydrogen/glucose ratio for several species and also depicts the effect of culturing type (batch or chemostat) and stirring. A crucial aspect of the metabolic shifts that occur in these microorganisms, is the way reducing equivalents are disposed off, or in other words how electron carriers like NAD(P) and ferredoxin are recycled. Since the pioneering studies of Jungermann, Decker and Thauer, many studies focussed on this aspect of product formation. Despite the large number of papers describing the metabolic shifts and the oxidoreductases involved, up to now no NADH: ferredoxin reductase or ferredoxin:NAD reductase has been purified and characterized. Although, the genome of several clostridia has been sequenced, it is not yet clear which genes code for these oxidoreductases.

Regulation of the redox metabolism is complex. Biochemical studies have shown that the acetyl-CoA/free CoA ratio determines the activity of the NADH: ferredoxin oxidoreductase.[6,46] In addition, the NADH/NAD ratio plays a role in regulating the electron flow.[65] Moreover, also the absolute level of NAD(H) may change substantially, depending on the growth rate or growth phase.[45,66] Also the H_2 concentration influences the flow of electrons from reduced ferredoxin to NAD and *vice versa*,[22] although the mechanism behind this is not clear. Unfortunately, because the specific enzymes and the corresponding genes have not been identified, transcriptional control of the electron flow has not been investigated, as yet.

11.7   The Genus *Caldicellulosiruptor*

The thermophilic organisms that are able to degrade cellulose can be divided into two categories; the thermophilic sporogenous clostridia, derived from compost piles or non-thermophilic sources (e.g. *Cl. thermocellum*), and the extremely thermophilic asporogenous microorganisms, isolated from thermal springs.[67] Several typical examples of the latter group can be found within the genus *Caldicellulosiruptor*, and because of their high capacity to produce hydrogen from biomass, we treat them here separately. The genus *Caldicellulosiruptor* contains as yet 5 validly described species which are all extremely thermophilic, Gram$^+$, and all but one are cellulolytic.[68] The ability to degrade cellulose at extremely thermophilic conditions is rather unusual and the highest temperature at which growth on cellulose has been described is 78 °C (record is set by *Caldicellulosiruptor kristjanssonii*.[69] The major fermentation end products are acetic acid, lactic acid, and ethanol in addition to hydrogen and carbon dioxide. The amount of lactate and ethanol is usually rather low, and concomitantly hydrogen levels are relatively high. For this reason and the cellulolytic capacity, *Ca. saccharolyticus* was selected for further research on biohydrogen production.[5,70] Moreover, *Ca. saccharolyticus* was selected by the DOE for genome sequencing, which is expected to be finished early 2007. *Ca. saccharolyticus* efficiently converts a wide range of biomass components viz. cellulose (Avicel, amorph), hemicellulose, pectin, α-glucans (starch, glycogen, pullulan), β-glucans (lichenan, laminarin), guar gum and gum locust bean resulting in the formation of hydrogen.[67]

Batch incubations (pH-controlled) resulted in hydrogen/glucose ratios of 3.3.[5,71] Relatively high $P(H_2)$ were found to initiate a shift to lactate production ($\sim 2 \times 10^4$ Pa) or to completely inhibit growth ($\sim 6 \times 10^4$ Pa).[70] However, continuous culture cultivation on glucose or xylose combined with hydrogen removal by nitrogen purging, resulted in complete fermentation to acetate, without formation of lactic acid or ethanol. Under these conditions the hydrogen acetate ratio comes close to the theoretical maximum of 4 (H. Goorissen et al., unpublished results). Moreover, preliminary fermentation data showed that glucose and xylose can be utilized simultaneously, and that catabolite repression by glucose does not occur.[72] From the viewpoint of hydrogen formation from biomass this is an important phenomenon, which enables efficient conversion of both cellulose and hemicellulose. *Ca. saccharolyticus* has been tested for hydrogen formation on various biomass hydrolysates (Paper sludge, Miscanthus, potato steam peels, Sweet Sorghum).[73] *Ca. saccharolyticus* displayed efficient behavior, producing 3-4 hydrogen per glucose equivalent. *Ca. saccharolyticus* has a preference for xylose if grown in mixtures with glucose (Figure 11.3a and 11.3b), although this xylose repression is concentration dependent. At a high initial xylose concentration xylose acted as a repressor. At a low initial xylose concentration, glucose repressed xylose utilization. Similar results were obtained for other pentose/hexose mixtures (results not shown), although some small differences in lactate production were observed.

The available genome sequence will enable detailed analysis of the catabolic and redox-mediating pathways by functional genomics. As soon as a genetic system can be developed for this organism, metabolic engineering may be used to improve the hydrogen generating system.

Ca. saccharolyticus uses the EM pathway for its glucose catabolism which suggests that reducing equivalents are produced as NADH and reduced ferredoxin.[71] Hydrogen formation may proceed by an NADH-dependent cytoplasmic Fe-only hydrogenase and a

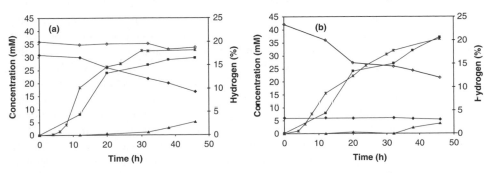

Figure 11.3 Growth of Ca. saccharolyticus *at high (a) and low (b) xylose concentrations in mixtures of glucose and xylose. Symbols:* ∗, *hydrogen;* ◊, *glucose;* ◆, *xylose;* ■, *acetate;* ▲, *lactate. (72)*

ferredoxin-dependent membrane-bound NiFe hydrogenase. The corresponding genes could be identified in the draft genome sequence of *Ca. saccharolyticus*, based on homology to the *Ta. tengcongensis* genes.[28] An NADH: ferredoxin reductase could not be found.

11.8   The Genus *Thermoanaerobacter*

The genus *Thermoanaerobacter* comprises a group of extreme thermophilic obligate anaerobes that ferment sugars to acetate, lactate, ethanol, CO_2 and hydrogen in different amounts. About 16 valid species have been recognized and they differ from *Caldicellulosiruptor* species in that they are able to respire with thiosulfate or sulfur, do not utilize cellulose, and may form endospores.[74] Some species are able to oxidize H_2 using thiosulfate or Fe(III) as electron- acceptor.[75,76] *Thermoanaerobacter ethanolicus* has been studied for its capacity to produce substantial amounts of ethanol, especially from pentoses.[77,78] In addition to ethanol also lactate is produced as reduced end product by several *Thermoanaerobacter* species.[27] A similar fermentation pattern was described for *Ta. tengcongensis*.[27] In correspondence with the formation of ethanol and lactate, hydrogen levels are usually low (Table 11.2). Despite the low hydrogen production by *Thermoanaerobacter* species under normal batch conditions, high levels (up to 4 H_2/glucose) can be attained in N_2-flushed fermentor systems, as was described for *Ta. tengcongensis*.[28]

Two different types of hydrogenases are responsible for H_2 formation in this organism: a ferredoxin dependent [NiFe]-hydrogenase and an NADH-dependent Fe-only hydrogenase. The former is membrane-bound and related to the energy-converting hydrogenase (Ech) from *Methanosarcina barkeri*, whereas the latter is located in the cytoplasm and contains FMN. The presence of an NADH-dependent hydrogenase in addition to a ferredoxin-dependent hydrogenase is unusual. This suggests that NADH can be directly used for hydrogen formation without the intermediate involvement of a NADH- ferredoxin reductase (oxidoreductase) as described for certain clostridia.[43,44] Culturing under a high $P(H_2)$ resulted in the formation of ethanol, and was accompanied by lower levels of the NADH-dependent hydrogenase and higher levels of NADPH-dependent alcohol- and acetaldehyde dehydrogenases.[28] These observations suggest that the activity of the NADH-dependent hydrogenase is regulated by the $P(H_2)$ and that under conditions of high $P(H_2)$ NADH is not used for H_2 formation but recycled via the formation of ethanol. However, the preference of the ethanol-forming enzymes for NADPH instead of NADH implies the involvement of an NADH:NADP transferase. Homologs of the NADH-dependent hydrogenase encoding genes (*hydA, hydB, hydC, hydD*) can be found in the genome of *Th ethanolicus*, but also in *Ca. saccharolyticus*, *Cl. thermocellum*, *Cl. phytofermentans* and *Tt. maritima*, suggesting that these microorganisms all are able to produce hydrogen directly from NADH.

The membrane-bound [NiFe]-hydrogenase of *Ta. tengcongsis* has been purified and appeared to consist of six polypeptides. The encoding genes were identified (*echABCDEF*) and shown to be clustered and followed by a gene cluster required for the biosynthesis of the [NiFe] center.[28] It was proposed that the membrane-bound Ech hydrogenase of *Ta. tengcongensis* may be able to generate a pmf and thus conserve energy by 'proton respiration' as was described for the Mbh hydrogenase of *P. furiosus*.[30] Comparable genomics revealed that the Ech gene cluster is present also in the genomes of *Cl. thermocellum*, *Cl. phytofermentans*, *Ca. saccharolyticus*, but not in *Tt. maritima* or the closely related *Ta.*

ethanolicus. It has recently been argued that the *ech* operon is an example of horizontal gene transfer, coming from the *Methanosarcina* genus or related species.[79]

11.9 The Genus *Thermotoga*

Members of the order *Thermotogales* are Gram⁻, anaerobic, thermophilic, and heterotrophic bacteria characterized by the presence of a toga, an outer sheath-like structure surrounding the cells. All members of this order (*Fervidobacterium, Thermotoga, Geotoga, Petrotoga, Marinitoga* and *Thermosipho*) are able to utilize complex carbohydrates and proteins, and all produce hydrogen to some extent. Sulfur compounds (thiosulfate, $S°$) can be used as electron acceptors. The genus *Thermotoga* has been investigated in more detail and for some species high levels of hydrogen production have been observed. Both *Tt. maritima* and *Tt. neapolitana* have been reported to produce up to 4 hydrogen per glucose.[26,80] Apparently these species are capable of directing all reducing equivalents towards proton reduction. *Tt. maritima* harbors a cytoplasmic Fe-only hydrogenase which contains a flavin binding site, suggesting that it may use NAD(P)H as electron donor.[81,82] Moreover, cell extracts of *Tt. maritima* catalyzed H_2-dependent NAD reduction. However, although the purified enzyme showed hydrogenase activity using the artificial electron carriers methyl viologen- and AQDS, it did not catalyze NAD(P)H-dependent H_2 formation or the reverse reaction.[82] *Cl. pasteurianum* contains a similar Fe-only hydrogenase which is ferredoxin-dependent, but which does not contain the flavin binding domain.[42] However, ferredoxin could not act as electron donor for the *Tt. maritima* hydrogenase and it is therefore assumed that the physiological role is the oxidation of NADH. Moreover, a homologous Fe-only hydrogenase was purified from *Ta. tengcongensis* which did show NADH-dependent hydrogenase activity (see above).

Tt. maritima also produces reduced ferredoxin in the pyruvate: ferredoxin oxidoreductase reaction, which is commonly recycled by proton reduction in a [NiFe]-hydrogenase. Such a ferredoxin-dependent hydrogenase complex (*ech* operon) has been purified from *Ta. tengcongensis* and the corresponding gene cluster has been identified.[28] Although, a similar hydrogenase cluster can be found in *Cl. thermocellum, Cl. phytofermentans*, and *Ca. saccharolyticus*, it is absent in *Tt. maritima*. *Tt. maritima*, on the other hand, contains another gene cluster (*Mbx* operon; 13 genes), which is highly homologous to gene clusters in the hyperthermophilic archaea *P. furiosus, P. abyssi, P. horikoschii* and *Thermococcus kodakaraensis*. These gene clusters are suggested to code for an energy transducing hydrogenase complex.[30,83] Whether the *mbx* operon in *Tt. maritima* also codes for such an energy transducing and ferredoxin-dependent hydrogenase, has not been experimentally confirmed.

Biochemical analyses and genomics of *Tt. maritima* have revealed that it uses the classical EM-pathway for glucose catabolism and that reducing equivalents are produced as NADH (glyceraldehyde-3-P dehydrogenase) and ferredoxin (pyruvate:ferredoxin oxidoreductase). *Tt. maritima* also contains the classical Entner-Doudoroff (ED) pathway which is apparently used alongside the EM pathway in a ratio of 15 %/85 %, respectively.[84] Whether the ED pathway may become dominant under certain conditions is not known. In the upper part of the ED pathway reducing equivalents are produced as NADPH,[85] which may only be used for anabolism, and thus are not available for hydrogen production.

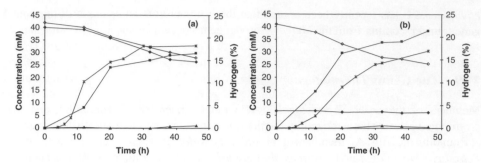

Figure 11.4 *Growth of* Tt. elfii *at high (a) and low (b) xylose concentrations in mixtures of glucose and xylose. Symbols:* ∗ *hydrogen;* ◇, *glucose;* ◆, *xylose;* ■, *acetate;* ▲, *lactate. (72)*

Several *Thermotoga* species are able to grow on complex carbohydrates, like cellulose or xylan and various endocellulases, cellobiohydrolases and xylanases have been purified and characterized.[86] The broad hydrolytic capacity combined with the high hydrogen producing potential makes the *Thermotoga* species good candidates for hydrogen production from biomass, although some species need 'rich' medium and high salt. The production of hydrogen by *Tt. neapolitana*, *Tt. elfii* and *Tt. maritima* on different biomass types, like sorghum and miscanthus, has been investigated (24). In general hydrolysates were not toxic and high levels of hydrogen were reached. Experiments with *Tt. elfii* (Figure 11.4a and 11.4b) revealed that xylose and glucose were used simultaneously if the xylose concentration was high (~40 mM). At low xylose concentration (~ 5 mM) glucose repressed xylose utilization similar to the pattern observed for *Ca. saccharolyticus*, although no lactate production was found.

11.10   The Genus *Pyrococcus/Thermococcus*

The genera with the highest temperature optimum that will be discussed here for hydrogen production are *Pyrococcus* and *Thermococcus*. These closely related genera represent a group of hyperthermophilic sulfur-reducing heterotrophs and the only archaea considered here. Various sugar- or amino acid based oligomers and polymers can be fermented by the different species, however, cellulose and xylan cannot be used. Most data on hydrogen production are obtained for *P. furiosus*, which is one of the best studied hyperthermophiles. As research during the past decade has shown, many archaea possess non-canonical versions of the glycolytic pathways.[87] Although *Pyrococcus* and *Thermococcus* species appeared to have an EM-like pathway,[88] various steps in the pathway were modified compared to the classical version.[87] Most importantly, the oxidation of glyceraldehyde-3-phosphate is not NAD-dependent but uses ferredoxin instead.[89] Moreover, the oxidation of GAP to 3-phosphoglycerate was not dependent on P_i and did not involve the intermediate formation of 1,3-bisphophoglycerate, and therefore it does not yield ATP by substrate-level phosphorylation.[90] At the level of pyruvate a second oxidation step occurs which is similar to the classical pyruvate:ferredoxin-oxidoreductase observed in e.g. clostridia. Thus, the main difference between the glycolytic pathways of *Pyrococcus/Thermoccus* and all other hydrogenic species is that all reducing equivalents are released as ferredoxin. Since hydrogen formation from ferredoxin is thermodynamically more favorable (*vide infra*), *Pyrococcus* should be able to produce hydrogen more easily. However, batch culturing has shown that in addition to acetate, CO_2 and alanine, *P. furiosus* produces ~2.6–3.0 H_2 per glucose equivalent,[29] which is not superior to the mesophiles, and even worse compared to *Tt. maritima*. Nevertheless, the amount of hydrogen that is produced is strongly influenced by the $P(H_2)$.[29] At low $P(H_2)$ more hydrogen (3.7 H_2/glucose) is produced as was shown by coculturing of *P. furiosus* with a hyperthermophilic methanogen. Likewise, in substrate-limited chemostat cultures, during N_2-flushing, hydrogen levels amounted to 3.9 H_2 per glucose (Tuininga and Kengen unpublished results).

The enzyme involved in transfer of the electrons (ferredoxin) to hydrogen has been investigated in detail for *P. furiosus*.[30,83] *P. furiosus* contains two cytoplasmic [NiFe]-containing hydrogenases (Sulfhydrogenase I and II) that are capable of reducing protons to hydrogen but also to reduce polysulfide to H_2S (37, 91) (Table 11.3). However, NADPH is the preferred electron donor for these bifunctional enzymes and not ferredoxin. A third cytopasmic enzyme has been described, which contains Fe-S centers and a flavin and which is also capable of polysulfide reduction. In addition, this sulfide dehydrogenase was shown to have ferredoxin:NADP oxidoreductase activity, and it was therefore initially hypothesized that reductant is transferred via ferredoxin to NADP and subsequently to hydrogen.[47] However, Silva et al. showed that the turnover of the sulfhydrogenase is too low to account for the flux of reductant from the glycolysis. Moreover, they could demonstrate that *P. furiosus* contains another membrane-bound [Ni-Fe] hydrogenase complex which can accept electrons from ferredoxin directly, without the involvement of NADP(H). This hydrogenase complex is encoded by the *mbh* operon, consisting of 14 ORFs. This hydrogenase was found to be unique in that it is rather resistant to inhibition by CO and that it exhibits an extremely high H_2 evolution to H_2 uptake activity ratio compared to other hydrogenases.[83] Moreover, it was shown that the enzyme could act as a proton pump, generating a proton-motive force, and coupled to the synthesis of ATP.[30] The genome contains a second gene cluster (*mbx* operon) coding for another [NiFe]-containing hydrogenase complex, and highly homologous to the *mbh* cluster (Table 11.3). Its physiological role is as yet unknown.

The ferredoxin:NADP oxidoreductase (or sulfide dehydrogenase) can become active under conditions of high $P(H_2)$, by transferring electrons from ferredoxin to polysulfide or to NADP. Subsequently, NADPH can be reoxidized by producing alanine from pyruvate using glutamate dehydrogenase (NADP-dependent) and alanine aminotransferase.[92]

11.11   Approaches for Improving Hydrogen Production

Ways to improve fermentative hydrogen production have been reviewed by Nath and Das.[93] Improvements can be made by the proper choice of the fermentative bacterium, by applying genetic modification to redirect biochemical pathways in such a way that less side products are formed and by a proper process design and process operation (gas sparging, creating a large biofilm surface, etc.). All these measures will optimize hydrogen formation only to maximally 4 molecules of hydrogen per molecule of glucose and may result in an increase of the hydrogen production rate. However, the challenge is to extend the yield to more than 4 hydrogen per glucose. In principle this might be possible by blocking certain genes of the glycolysis and expressing genes that encode for enzymes of the pentose phosphate pathway. While under fermentative conditions the EM pathway can produce up to 4 H_2/glucose (concomitant with 2 acetate), the PP cycle can theoretically produce 12 H_2/glucose because glucose can be entirely oxidized to CO_2. This was demonstrated in an *in vitro* system using the enzymes of the oxidative pentose phosphate cycle.[94,95] By combining the enzymes of the PP cycle (from Yeast) with the NADP-dependent sulfhydrogenase of *P. furiosus*, in an in vitro system,11.6 mol of hydrogen could be formed per mol of glucose-6-P. However, the instability of some of the enzymes and of NADP(H) hampers the practical application of the method. To overcome this problem some oxidative PP cycle enzymes from *Tt. maritima* were cloned into *E. coli*,[96] but an efficient hydrogen producing system has not been obtained yet. The genome sequences of *Tt. maritima*, *P. furiosus* and *Caldicellusiruptor saccharolyticus* are available now, enabling genetic engineering approaches within these promising hydrogen-producing microorganisms, provided that genetic systems can be developed.

A main breakthrough would be achieved when acetate would be efficiently converted to carbon dioxide and hydrogen. Thermophilic bacteria, like *Thermoacetogenium phaeum* and *Clostridium ultenense*[97,98] have the biochemical potential to do so. At low $P(H_2)$, these bacteria employ the acetyl-CoA cleavage pathway to convert acetate to carbon dioxide and hydrogen. They even can grow by acetate conversion. In principle also homoacetogenic bacteria and enterobacteria have the biochemical potential to oxidize acetate and form hydrogen, the former via the acetyl-CoA cleavage pathway and the latter via the citric acid cycle. However, energy input is needed to enable hydrogen production to high levels. One way to introduce extra energy is light. Light energy enables phototrophic bacteria of the genus *Rhodopseudomonas* to produce high levels of hydrogen from acetate. Such an integrated dark-light fermentation is under investigation[99] and its perspectives have been discussed by Nath and Das.[93]

Another way to introduce energy in a dark fermentation is via electricity.[100] The concept of electricity-mediated electrolysis of organic compounds was introduced by Liu et al.[101] and Rozendal et al.[102] They showed that in a biofuel cell acetate can yield hydrogen and carbon dioxide and that this type of hydrogen formation is much more cost-effective than electricity mediated electrolysis of water to form hydrogen and oxygen. A rough calculation showed that the equivalent of one hydrogen is needed to produce the electricity for the electrolysis of acetate. This implies that (a) a net yield of 10 molecules of hydrogen per molecule of glucose can be obtained, and (b) hydrogen formation is not restricted to (poly)saccharides as substrates, but that all kinds of organic (waste) components are feasible substrates.

11.12 Concluding Remarks

Dark hydrogen formation is performed by many anaerobic and facultative anaerobic microorganisms, which differ, however, in the amount of hydrogen that is produced per glucose. Thermophiles and extreme thermophiles appear to be superior in this respect, as the amount of hydrogen approaches the apparent maximum of 4 H_2/glucose and less side-products are produced. These observations are supported by the thermodynamics of the H_2-producing reactions. Dependent on the electron carrier used, energy-conserving membrane-bound hydrogenases as well as cytoplasmic hydrogenases appear to catalyze the ultimate H_2 forming reactions. However, to make biohydrogen formation feasible we need to increase the H_2/glucose ratio beyond the current maximum of 4. To achieve this, future research will focus on metabolic engineering to decrease the amount of acetate or on acetate oxidation by electricity-mediated electrolysis or photofermentation.

Acknowledgment

Our research on biological hydrogen formation was financially supported by the Dutch Programme Economy, Ecology, Technology (EET), a joint initiative of the Ministries of Economic Affairs, Education, Culture and Sciences, and of Housing, Spatial Planning and the Environment (EETK03028 BWPII) and the Commission of the European Communities, Sixth Framework Programme, Priority 6, Sustainable Energy Systems (019825 HYVOLUTION).

References

1. B. Schink, and A. J. M. Stams, Syntrophism among prokaryotes, *in The prokaryotes* (*electronic third edition*), M. Dworkin, K.-H. Schleifer and E. Stackebrandt (Eds), Springer Verlag, New York, 2002.
2. S. H. Zinder, Syntrophic acetate oxidation and 'reversible acetogenesis', in *Acetogenesis*, H. L. Drake (Ed), Chapman and Hall, New York, pp 386–415, 1994.
3. I. K. Kapdan and F. Kargi, Bio-hydrogen production from waste materials, *Enz. Microbiol. Technol.*, **38**, 569–582 (2006).
4. P. A. M. Claassen, J. B. van, A. M. Lopez Contreras, E. W. J. van Niel, L. Sijtsma, A. J. M. Stams, S. S. de Vries and R. A. Weusthuis, Utlisation of biomass for the supply of energy carriers, *Appl. Microbiol. Biotechnol.*, **52**, 741–755 (1999).
5. E. W. J. van Niel, M. A. W. Budde, G.G. de Haas, F. J. Van Der Wal, P. A. M. Claassen and A. J. M. Stams, Distinctive properties of high hydrogen producing extreme thermophiles, *Caldicellulosiruptor saccharolyticus and Thermotoga elfii*, *Int. J. Hydrogen*, **27**, 1391–1398 (2002).
6. R. K. Thauer, K. Jungermann and K. Decker, Energy conservation in chemotrophic anaerobic bacteria, *Bacteriol. Rev.*, **41**, 100–180 (1977).
7. J. P. Amend and E. L. Shock, Energetics of overall metabolic reactions of thermophilic and hyperthermophilic Archaea and Bacteria, *FEMS microbiol. Rev.*, **25**, 175–243 (2001).
8. J. P. Amend and A. V. Plyasunov, Carbohydrates in the thermophilic metabolism: calculation of the standard molal thermodynamic properties of aqueous pentoses and hexoses at elevated temperatures and pressures, *Geochim. Cosmochim. Acta*, **64**, 3901–3917 (2001).
9. T. de Vrije and P. A. M. Claassen, Dark hydrogen fermentations, in *Bio-methane & Biohydrogen*, J. H. Reith, R. H. Weiffels and H. Barten (eds.), Smiet Offset, The Hague, pp 103–123, 2003.
10. S. Tanisho, Y. Suzuki and N. Wakao, Fermentative hydrogen evolution by *Enterobacter aerogenes* strain E82005, *Int. J. Hydrogen Energy*, **12**, 623–627 (1987).

11. S. Tanisho and Y. Ishiwata, Continuous hydrogen production from molasses by the bacterium *Enterobacter aerogenes. Int. J. Hydrogen Energy*, **19**, 807–812 (1994).
12. M. A. Rachman, Y. Nakashimada, T. Kakizono and N. Nishio, Hydrogen production with high yield and high evolution rate by self-flocculated cells of *Enterobacter aerogenes* in a packed-bed reactor. *Appl. Microbiol. Biotechnol.*, **49**, 450–454 (1998).
13. L. Minnan, H. Jinli, W. Xiaobin, X. Huijuan, C. Jinzao, L. Chuannan, Z. Fengzhang and X. Liangshu, Isolation and characterization of a high H_2-producing strain *Klebsiella oxytoca* HP1 from a hot spring, *Res. Microbiol.*, **156**, 76–81 (2005).
14. J. G. van Andel, G. R. Zoutberg, P. M. Crabbendam, and A. M. Breure, Glucose fermentation by *Clostridium butyricum* grown under a self generated gas atmosphere in chemostat culture, *Appl. Microbiol. Biotechnol.*, **23**, 21–26 (1985).
15. B. H. Kim, P. Bellows, R. Datta and J. G. Zeikus, Control of carbon and electron flow in *Clostridium acetobutylicum* fermentations: utilization of carbon monoxide to inhibit hydrogen production and to enhance butanol yields, *Appl. Environ. Microbiol.*, **48**, 764–770 (1984).
16. F. Taguchi, N. Mizukami, K. Hasegawa and T. Saito-Taki, Microbial conversion of arabinose and xylose to hydrogen by a newly isolated *Clostridium* sp. No.2, *Can. J. Microbiol.*, **40**, 228–233 (1994).
17. S. G. Pavlostathis, T. L. Miller and M. J. Wolin, Fermentation of insoluble cellulose by continuous cultures of *Ruminococcus albus*, *Appl. Environ. Microbiol.*, **54**, 2655–2659 (1988).
18. S. G. Pavlostathis, T. L. Miller, and M. J. Wolin, Cellulose fermentation by cultures of *Ruminococcus albus* and *Methanobrevibacter smithii*, *Appl. Microbiol. Biotechnol.*, **33**, 109–116 (1990).
19. M. Vancanneyt, P. de Vos, M. Maras and J. De Ley, Ethanol production in batch and continuous culture from some carbohydrates with *Clostridium thermosaccharolyticum* LMG 6564, *Syst. Appl. Microbiol.*, **13**, 382–387 (1990).
20. D. Freier, C. P. Mothershed and J. Wiegel, Characterisation of *Clostridium thermocellum* JW 20, *Appl. Environ. Microbiol.*, **54**, 204–211 (1988).
21. G. D. Bothun, J. A. Berberich, B. L. Knutson, H. J. Strobel and S. E. Nokes, Metabolic selectivity and growth of *Clostridium thermocellum* in continuous culture under elevated hydrostatic pressure. *Appl. Microbiol. Biotechnol.*, **65**, 149–157 (2004).
22. R. J. Lamed, J. H. Lobos and T. M. Su, Effects of stirring and hydrogen on fermentation products of *Clostridium thermocellum*, *Appl. Environ. Microbiol.*, **54**, 1216–1221 (1988).
23. R. Islam, N. Cicek, R. Sparling and D. Levin, Effect of substrate loading on hydrogen production during anaerobic fermentation by *Clostridium thermocellum* 27405, *Appl. Microbiol. Biotechnol.*, **72**, 576–83 (2006).
24. T. de Vrije, G.G. de Haas, G.B. Tan, E.R.P. Keijsers and P.A.M. Claassen, Pretreatment of Miscanthus for hydrogen production by *Thermotoga elfii. Int. J. Hydrogen Energy*, **27**, 1381–1390 (2002).
25. S. A. Van Ooteghem, A. Jones, D. van der Lelie, B. Dong and D. Mahajan, H_2 production and carbon utilization by *Thermotoga neapolitana* under anaerobic and microaerobic growth conditions, *Biotechnol. Lett.*, **26**, 1223–1232 (2004).
26. C. Schröder, M. Selig and P. Schönheit, Glucose fermentation to acetate, CO2 and H2 in the anaerobic hyperthermophilic eubacterium *Thermotoga maritima* involvement of the Embden-Meyerhof pathway. *Arch. Microbiol.*, **161**, 460–470 (1994).
27. Y. Xue, Y. Xu, Y. Liu, Y. Ma and P. Zhou P, *Thermoanaerobacter tengcongensis sp.* a novel anaerobic, saccharolytic, thermophilic bacterium isolated from a hot spring in Tengcong, China, *Int J Syst Evol Microbiol*, **51**, 1335–1341 (2001).
28. B. Soboh, D. Linder and R. Hedderich, A multisubunit membrane-bound [NiFe] hydrogenase and an NADH-dependent Fe-only hydrogenase in the fermenting bacterium *Thermoanaerobacter tengcongensis*, *Microbiology* **150**, 2451–63 (2004).

29. S. W. M. Kengen and A. J. M. Stams, Growth and energy conservation in batch cultures of *Pyrococcus furiosus*, *FEMS Microbiol. Lett.*, **117**, 305–310 (1994).

30. R. Sapra, K. Bagramyan and M. W. W. Adams, A simple energy conserving system: proton reduction coupled to proton translocation, *Proc. Natl. Acad. Sci. USA*, **100**, 7545–7550 (2003).

31. R. N. Schicho, K. Ma, M. W. W. Adams, and R. M. Kelly, Bioenergetics of sulfur reduction in the hyperthermophilic archaeon *Pyrococcus furiosus*, *J. Bacteriol.*, **175**, 1823–1830 (1993).

32. T. Kanai, H. Imanaka, A. Nakajima, K. Uwamori, Y. Omori, T. Fukui, H. Atomi and T. Imanaka, Continuous hydrogen production by the hyperthermophilic archaeon, *Thermococcus kodakaraensis* KOD1, *J. Biotechnol.*, **116**, 271–282 (2005).

33. A. J. M. Stams, Metabolic interactions between anaerobic bacteria in methanogenic environments, *Antonie van Leeuwenhoek*, **66**, 271–294 (1994).

34. G. D. Watt and A. Burns, The thermodynamical characterization of sodium dithionite, flavin mononucleotide, flavin-adenine dinucleotide and methyl and benzyl viologens as low-potential reductants for biological systems, *Biochem. J.*, **152**, 33–37 (1975).

35. K. Burton, The enthalpy change for the reduction of nicotinamide-adenine denucleotide, *Biochem. J.*, **143**, 365–368 (1974).

36. C. D. Boiangiu, E. Jayamani, D. Brügel, G. Herrmann, J. Kim, L. Forzi, R. Hedderich, I. Vgenopoulou, A. J. Pierik, J. Steuber, and W. Buckel, Sodium ion pumps and hydrogen production in glutamate fermenting anaerobic bacteria, *J. Mol. Microbiol. Biotechnol.*, **10**, 105–119 (2005).

37. K. Ma, R. N. Schicho, R. M. Kelly and M. W. W. Adams, Hydrogenase of the hyperthermophile *Pyrococcus furiosus* is an elemental sulfur reductase or sulfhydrogenase: Evidence for a sulfur-reducing hydrogenase ancestor, *Proc. Natl. Acad. Sci. USA*, **90**, 5341–5344 (1993).

38. M. W. W. Adams, The structure and mechanism of iron-hydrogenases, *Biochim. Biophys. Acta*, **1020**, 115–145 (1990).

39. A. E. Przybyla, J. Robbins, N. Menon and H. D. Peck Jr., Structure-function relationships among the nickel-containing hydrogenases, *FEMS Microbiol. Rev.* **88**, 109–136 (1992).

40. S. P. J. Albracht, Nickel hydrogenases: in search of the active site. *Biochim. Biophys. Acta* **1188**, 167–204 (1994).

41. R. Hedderich and L. Forzi, Energy-converting [NiFe] hydrogenases: more than just H_2 activation, *J. Mol. Microbiol. Biotechnol.*, **10**, 92–104 (2005).

42. M. W. W. Adams, E. Eccleston and J. B. Howard, Iron-sulfur clusters of hydrogenase I and hydrogenase II of *Clostridium pasteurianum*, *Proc. Natl. Acad. Sci. USA*, **86**, 4932–4936 (1989).

43. K. Jungermann, G. Leimenstoll, E. Rupprecht and R. K. Thauer, Demonstration of NADH-ferredoxin reductase in two saccharolytic clostridia, *Arch. Microbiol.*, **80**, 370–372 (1971).

44. G. Gottschalk and A. A. Chowdhury, Pyruvate synthesis from acetyl coenzyme A and carbon dioxide with NADH(2) or NADPH(2) as electron donors, *FEBS Lett.* **2**, 342–344 (1969).

45. E. Guedon, S. Payot, M. Desvaux, and H. Petitdemange, Carbon and electron flow in *Clostridium cellulolyticum* grown in chemostat culture on synthetic medium, *J. Bacteriol.*, **181**, 3262–3269 (1999).

46. S. Saint-Amans, L. Girbal, J. Andrade, K. Ahrens and P. Soucaille, Regulation of carbon and electron flow in *Clostridium butyricum* VPI 3266 grown on glucose-glycerol mixtures, *J. Bacteriol.*, **183**, 1748–1754 (2001).

47. K. Ma and M. W. W. Adams, Sulfide dehydrogenase from the hyperthermophilic archaeon *Pyrococcus furiosus*: a new multifunctional enzyme involved in the reduction of elemental sulphur, *J. Bacteriol.*, **176**, 6509–6517 (1994).

48. W.A. Wood, Fermentation of carbohydrates and related compounds, *in The Bacteria 2*, I. C. Gunsales and R. Y. Stanier (Eds), Academic Press, New York, London, pp 59–149, 1961.

49. M. J. Axley, D. A. Grahame and T. C. Stadtman, *Escherichia coli* formate-hydrogen lyase. Purification and properties of the selenium-dependent formate dehydrogenase component, *J. Biol. Chem.*, **265**, 18213–18218 (1990).

50. R. G. Sawers, Formate and its role in hydrogen production in Escherichia coli, *Biochem. Soc. Trans.*, **33**, 42–46 (2005).

51. M. Hakobyan, H. Sargsyan and K. Bagramyan, Proton translocation coupled to formate oxidation in anaerobically grown fermenting *Escherichia coli*, *Biophys. Chem.*, **115**, 55–61 (2005).

52. A. Bisaillon, J. Turcot and P. C. Hallenbeck, The effect of nutrient limitation on hydrogen production by batch cultures of *Escherichia coli*, *Int. J. Hydrogen Energy*, **31**, 1504–1508 (2006).

53. N. Kumar, A. Ghosh and D. Das, Redirection of biochemical pathways for the enhancement of H$_2$ production by *Enterobacter cloacae*, *Biotechnol. Lett.*, **23**, 537–541 (2001).

54. K. Nath, A. Kumar and D. Das, Effect of some environmental parameters on fermentative hydrogen production by *Enterobacter cloacae* DM11, *Can. J. Microbiol.*, **52**, 525–532 (2006).

55. M. A. Rachman, Y. Furutani, Y. Nakashimada, T. Kakizono and N. Nishio, Enhanced hydrogen production in altered mixed acid fermentation of glucose by *Enterobacter aerogenes*, *J. Ferm. Bioeng.*, **83**, 358–363 (1997).

56. T. Ito, Y. Nakashimada, T. Kakizono and N. Nishio, High-yield production of hydrogen by *Enterobacter aerogenes* mutants with decreased α-acetolactate synthase activity, *J. Biosci. Bioengin.*, **97**, 227–232 (2004).

57. A. Yoshida, T. Nishimura, H. Kawaguchi, M. Inui and H. Yukawa, Enhanced hydrogen production from glucose using l*dh* – and *frd*-inactivated *Escherichia coli* strains, *Appl. Microbiol. Biotechnol.* **73**, 67–72 (2006).

58. A. Yoshida, T. Nishimura, H. Kawaguchi, M. Inui and H. Yukawa, Enhanced hydrogen production from formic acid by formate hydrogen lyase-overexpressing *Escherichia coli* strains, *Appl. Environ. Microbiol.*, **71**, 6762–6768 (2005).

59. F. Canganella and J. Wiegel, The potential of thermophilic clostridia in biotechnology, *in The clostridia and biotechnology*, D. R. Woods (ed.),. Butterworth-Heinemann, Boston, Mass., pp. 393–429, 1993.

60. L. H. Carreira and L. G. Ljungdahl, Production of ethanol from biomass using anaerobic thermophilic bacteria, *in Liquid fuel developments.*, D. L. Wise (ed.), CRC Press, Boca Raton, Fla., pp. 1–28, 1993.

61. G. M. Awang, G. A. Jones and W. M. Ingledew The acetone-butanol-ethanol fermentation, *Crit. Rev. Microbiol.*, **15** Suppl., 1, 33–67 (1988).

62. R. Lamed and G. Zeikus, Ethanol production by thermophilic bacteria: relationship between fermentation product yields of catabolic enzyme activities in *Clostridium thermocellum*, *J. Bacteriol.*, **144**, 569–578 (1980).

63. J. S. Chen and D. K. Blanchard, Purification and properties of the H2-oxidizing (uptake) hydrogenase of the N2-fixing anaerobe *Clostridium pasteurianum* W5, *Biochem. Biophys. Res. Commun.*, **122**, 9–16 (1984).

64. J. S. Chen and L. E. Mortenson, Purification and properties of hydrogenase from *Clostridium pasteurianum* W5, *Biochim. Biophys. Acta*, **371**, 283–298 (1974).

65. S. Payot, E. Guedon, E. Gelhaye and H. Petitdemange, Induction of lactate production associated with a decrease in NADH cell content enables growth resumption of *Clostridium cellulolyticum* in batch cultures on cellobiose, *Res. Microbiol.*, **150**, 465–473 (1999).

66. E. Guedon, S. Payot, M. Desvaux and H. Petitdemange, Relationships between cellobiose catabolism, enzyme levels, and metabolic intermediates in *Clostridium cellulolyticum* grown in a synthetic medium. *Biotechnol. Bioengin.*, **67**, 327–335 (2000).

67. F. A. Rainey, A. M. Donnison, P. H. Janssen, D. Saul, A. Rodrigo, P. L. Bergquist, R. M. Daniel, E. Stackebrandt and H. W. Morgan, Description of *Caldicellulosiruptor saccharolyticus* gen. nov., sp. nov: An obligately anaerobic extremely thermophilic, cellulolytic bacterium, *FEMS Microbiol. Lett.*, **120**, 263–266 (1994).

68. U. Onyenwoke, Y.-J. Lee, S. Dabrowski, B. K. Ahring and J. Wiegel, Reclassification of *Thermoanaerobium acetigenum* as *Caldicellulosiruptor acetigenus* comb. nov. and emendation of the genus description, *Int. J. Syst. Evol. Microbiol.*, **56**, 1391–1395 (2006).

69. S. Bredholt, J. Sonne-Hansen, P. Nielsen, I. M. Mathrani and B. K. Ahring, *Caldicellulosiruptor kristjanssonii* sp. nov., a cellulolytic, extremely thermophilic, anaerobic bacterium, *Int. J. Syst. Bacteriol.*, **49**, 991–996 (1999).

70. E. W. J. van Niel, P. A. M. Claassen and A. J. M. Stams, Substrate and product inhibition of hydrogen production by the extreme thermophile, *Caldicellulosiruptor saccharolyticus*, *Biotechnol. Bioeng.*, **81**, 255–262 (2003).

71. T. de Vrije, A. E. Mars, M. A. W. Budde, M. H. Lai, C. Dijkema, P. de Waard and P. A. M. Claassen, Glycolytic pathway and hydrogen yield studies of the extreme thermophile *Caldicellulosiruptor saccharolyticus*, *Appl. Microbiol. Cell Physiol.*, DOI 10.1007/s00253-006-0783-x (2007).

72. H. P. Goorissen and A. J. M. Stams, Biological hydrogen production by moderately thermophilic anaerobic bacteria. Proceedings of 16th World Hydrogen Energy Conference, june 13-16, Lyon, France, 2006.

73. Z. Kádár, T. de Vrije, G. E. van Noorden, M. A. W. Budde, Z. Szengyel, K. Réczey and P. A. M. Claassen, Yields from glucose, xylose, and paper sludge hydrolysate during hydrogen production by the extreme thermophile *Caldicellulosiruptor saccharolyticus*, *Appl. Biochem. Biotechnol.*, **113-116**, 497–508 (2004).

74. I. V. SubbotinaI, N. A. Chernyh, T. G. Sokolova, I. V. Kublanov, E. A. Bonch-Osmolovskaya and A. V. Lebedinsky, Oligonucleotide probes for the detection of representatives of the genus *Thermoanaerobacter*, *Microbiology* **72**, 331–339 (2003).

75. A. I. Slobodkin, T. P. Tourova, B. B. Kuznetsov, N. A. Kostrikina, N. A. Chernyh and E. A. Bonch-Osmolovskaya, *Thermoanaerobacter siderophilus* sp. nov., a novel dissimilatory Fe(III)-reducing, anaerobic, thermophilic bacterium, *Int. J. Syst. Bacteriol.*, **49**, 1471–1478 (1999).

76. M.-L. Fardeau, J.-L.Cayol, M. Magot and B. Ollivier, H2 oxidation in the presence of thiosulfate by a *Thermoanaerobacter* strain isolated from an oil-producing well, *FEMS Microbiol. Lett.* **113**, 327–332 (1993).

77. L. S. Lacis and H. G. Lawford, Ethanol production from xylose by Thermoanaerobacter ethanolicus in batch and continuous culture, *Arch. Microbiol.*, **150**, 48–55 (1988).

78. L. S. Lacis and H. G. Lawford, *Thermoanaerobacter ethanolicus* growth and product yield from elevated levels of xylose or glucose in continuous cultures, *Appl. Environ. Microbiol.*, **57**, 579–585 (1991).

79. A. Calteau, M. Gouy and G. Perriere, Horizontal transfer of two operons coding for hydrogenases between bacteria and archaea, *J. Mol. Evol.*, **60**, 557–565 (2005).

80. S. A. van Ooteghem , S. Beer and P. C. Yue, Hydrogen production by the thermophilic bacterium *Thermotoga neapolitana*, *Appl. Biochem. Biotechnol.*, **98–100**, 177–89 (2002).

81. M. F. J. M. Verhagen, T. O'Rourke and M. W. W. Adams, The hyperthermophilic bacterium, *Thermotoga maritima*, contains an unusually complex iron-hydrogenase: amino acid sequence analyses versus biochemical characterization, *Biochim. Biophys. Acta*, **1412**, 212–219 (1999).

82. M. F. J. M. Verhagen and M. W. W. Adams, Fe-only hydrogenase from *Thermotoga maritime*, *Method. Enzymol.*, **331**, 216–226 (2001).

83. P. J. Silva, E. C. Van Den Ban, H. Wassink, H. Haaker, B. de Castro, F. T. Robb, W. R. Hagen, Enzymes of hydrogen metabolism in *Pyrococcus furiosus*. *Eur. J. Biochem.*,. **267**, 6541–6551 (2000).

84. M. Selig, K. B. Xavier, H. Santos and P. Schonheit,.Comparative analysis of Embden-Meyerhof and Entner-Doudoroff glycolytic pathways in hyperthermophilic archaea and the bacterium *Thermotoga*, *Arch. Microbiol.*, **167**, 217–32 (1997).

85. T. Hansen, B. Schlichting and P. Schönheit, Glucose-6-phosphate dehydrogenase from the hyperthermophilic bacterium Thermotoga maritima: expression of the g6pd gene and characterization of an extremely thermophilic enzyme, *FEMS Microbiol. Lett.*, **216**, 249–53 (2002).

86. J. D. Bok, D. A. Yernool and D. E. Eveleigh, Purification, characterization, and molecular analysis of thermostable cellulases CelA and CelB from *Thermotoga neapolitana*, *Appl. Environ. Microbiol.*, **64**, 4774–4781 (1998).

87. C. H. Verhees, S. W. M. Kengen, J. E. Tuininga, G. J. Schut, M. W. W. Adams, W. M. de Vos and J. vander Oost, The unique features of glycolytic pathways in Archaea, *Biochem. J.*, **375**, 231–246 (2003).

88. S. W. M. Kengen, F. A. M. de Bok, N. D. van Loo, C. Dijkema, A. J. M. Stams and W. M. de Vos, Evidence for the operation of a novel Embden-Meyerhof pathway that involves ADP-dependent kinases during sugar fermentation by *Pyrococcus furiosus*, *J. Biol. Chem.*, **269**, 17537–17541 (1994).

89. S. Mukund and M. W. W. Adams, Glyceraldehyde-3-phosphate ferredoxin oxidoreductase, a novel tungsten-containing enzyme with a potential glycolytic role in the hyperthermophilic archaeon, *Pyrococcus furiosus*, *J. Biol. Chem.*, **270**, 8389–8392 (1995).

90. S. W. M. Kengen, A. J. M. Stams and W. M. de Vos, Sugar metabolism of hyperthermophiles. *FEMS Microbiol. Rev.*, **18**, 119–137 (1996).

91. K. Ma, R. Weiss and M. W. W. Adams, Characterization of sulfhydrogenase II from the hyperthermophilic Archaeon *Pyrococcus furiosus* and assessment of its role in sulfur reduction, *J. Bacteriol.*, **182**, 1864–1871 (2000).

92. S. W. M. Kengen and A. J. M. Stams, Formation of L-alanine as a reduced end product in carbohydrate fermentation by the hyperthermophilic archaeon *Pyrococcus furiosus*, *Arch. Microbiol.*, **161**, 168–175 (1994).

93. K. Nath and D. Das, Improvement of fermentative hydrogen production: various approaches, *Appl. Microbiol. Biotechnol.*, **65**, 520–529 (2004).

94. J. Woodward, S. M. Mattingly, M. Danson, D. Hough, N. Ward and M. Adams, In vitro hydrogen production by glucose dehydrogenase and hydrogenase. *Nature Biotechnol.*, **14**, 872–874 (1996).

95. J. Woodward, M. Orr, K. Cordray and E. Greenbaum, Enzymatic production of biohydrogen, *Nature*, **405**, 1014–1015 (2000).

96. J. Woodward, N. I. Heyer, J. P. Getty, H. M. O'Neill, E. Pinkhassik and B. R. Evans, Efficient hydrogen production using enzymes of the pentose phosphate pathway, *Proc. DOE Hydrog. Prog. Rev.* NREL/CP-610-32405 (2002).

97. S. Hattori, Y. Kamagata and H. Shoun, *Thermacetogenium phaeum* gen. nov., sp. nov., a strictly anaerobic, thermophilic syntrophic acetate-oxidizing bacterium, *Int. J. Syst. Evol. Microbiol.*, **50**, 1601–1605 (2000).

98. A. Schnürer, B. Schink and B. H. Svenson, *Clostridium ultunense* sp. nov., a mesophilic bacterium oxidizing acetate in syntrophic association with a hydrogenotrophic methanogenic bacterium, *Intl. J. Syst. Bacteriol.*, **46**, 1145–1152 (1996).

99. P. A. M. Claassen and T. de Vrije, Non-thermal production of pure hydrogen from biomass: HYVOLUTION, *Int. J. Hydrogen Energy*, **31**, 1416 – 1423 (2006).

100. B. E. Logan, B. Hamelers, R. Rozendal, U. Schröder, J. Keller, S. Freguia, P. Aelterman, W. Verstraete and K. Rabaey, Microbial fuel cells: methodology and technology, *Envir. Sci. Technol.*, **40**, 5181–5192 (2006).
101. H. Liu, S. Grot and B. E. Logan, Electrochemically assisted microbial production of hydrogen from acetate, *Env. Sci. Technol.*, **39** 4317–4320 (2005).
102. R. A. Rozendal, H. V. M. Hamelers, G. J. W. Euverink, S. J. Metz and C. Buisman, Principle and perspectives of hydrogen production through biocatalyzed electrolysis, *Int. J. Hydrogen Energy*, **31**, 1632–1640 (2006).

12

Improving Sustainability of the Corn-Ethanol Industry

Paul W. Gallagher

Department of Economics, Iowa State University, Iowa, USA

Hosein Shapouri

USDA, OCE, OE, Washington, DC, USA

12.1 Introduction

Two criteria based on characteristics of plant growth establish when bio-fuels can provide sustainable energy for society. The first criteria: enough solar energy stored during plant growth becomes available for man's use. Pimentel's early evaluation of the US ethanol industry calculated the ratio of BTUs in ethanol: BTUs from fossil energy of corn and ethanol production at less than one. He concluded that ethanol is not sustainable energy, and questioned the industry's existence. Recent energy balance ratios that include adjustments for co-product feed and higher energy efficiency in corn/ethanol production suggest a moderate contribution from captured solar energy. The ratio is around 1.3 (Shapouri et al., 2002). Dale questions the relevance of the net energy criteria, noting that economic value creation is consistent with energy ratios less than or near one. Dale (2007) proposes a second criteria for a sustainable fuel: enough CO_2 in the atmosphere is converted to carbon in the plant and O_2 in the atmosphere through photosynthesis and plant growth to improve global-warming. Comparing greenhouse gas emissions from a refinery and an ethanol plant, some have calculated that emissions could be about 20 % lower with to-day's corn-ethanol instead of the corresponding output of petroleum-based gasoline (Wang

Biofuels Edited by Wim Soetaert and Erick J. Vandamme

et al., 1999). Thus, recent calculations of energy ratios and CO_2 emission comparisons both suggest that the corn-ethanol industry is sustainable. Further, public policies to ensure the corn-ethanol industry's existence find moderate justification from both sustainability measures.

Analysis of corn-ethanol should now focus on incremental changes that could occur in the market economy and also improve sustainability. For instance, biomass power is economically competitive and potentially more sustainable than fossil energy inputs (Gallagher et al., 2006). We look at three possibilities for replacing fossil fuel processing energy with biomass: corn stover, willow, and distillers' grains. We also calculate energy ratio improvements. In our CO_2 analysis, we also take a broad view of the ethanol industry as joint producer of fuel and livestock feed in a market context. We show that using biomass power for processing energy, instead of coal and/or natural gas, can improve the sustainability of the corn-ethanol industry substantially. However, the best form of biomass power depends on the nature of the corn supply and demand adjustments that underlie ethanol production.

12.2   Energy Balance

A reduction in fossil energy inputs generally improves the energy balance ratio of biofuel output over external fuel inputs. Numerous energy efficiency improvements on the farm and in the ethanol plant have contributed to an increasingly favorable energy balance ratio for corn-ethanol. In the corn processing industry, dry mills typically use a combination of natural gas and market purchases of (coal-based) electricity. But energy balance ratios for sugar-ethanol are much higher, because crop residues are used for processing energy (Gallagher et al., 2006). Conceivably, energy balance could improve with a shift to biomass power.

Table 12.1 compares energy balance situation for a baseline fossil-fuel power and several alternative approaches to biomass power. In the first column, ethanol conversion energy reflects the heat content required for natural gas-based process heat (Shapouri et al., 2002, p. 9, Table 6). Also, market purchases of electricity are calculated from processing electricity requirements, an assumed 30 % efficiency for electrical power production, and the BTU content of the required coal. Three biomass power situations are shown in the next three columns: corn stover, willow, and distillers' grains. All of these biomass alternatives replace the natural gas and coal energy.

But the energy input investment for each biomass power source is unique.

For the willow input, the external energy input is the fuel for planting and maintaining the willow crop, including the energy embodied in machinery depreciation. Also, the energy expended in fertilizer production, distribution, and application is included. We use the estimates provided by Heller et al. (2003). Also, we have calculated the annual average energy use over the 23 year life of a willow plantation. A comparison of external processing energy for willow, 1,159BTU/gallon ethanol, and in the baseline, 48,771BTU/gallon, is quite dramatic – the difference is the sunlight energy stored and then used in biomass power generation.

It is the external energy investment in corn, already included in the baseline energy balance calculation, that also yields the other biomass components of corn stover and

Table 12.1 Corn ethanol dry mill: energy use and net energy value with conventional and biomass power for processing

| | Btu/gallon | | | |
Processing Power Configuration	Natural gas & purchased electricity	Biomass power: Corn Stover	Biomass power: willow (SRWC)	Distillers' grain for power
Corn production	21803	21803	21803	21803
Corn transport	2284	2284	2284	2284
Ethanol conversion	48771[1]	1687[2]	1159[3]	0[5]
Ethanol distribution	1588	1588	1588	1588
Total energy used	74446	27362	26834.3	25675
By-product Credit	13115[4]	13115	13115	0
Energy used, net of byproduct credit	61331	14247	13719	25675
Ethanol energy produced	84530	84530	84530	84530
Energy ratio, w/o byproduct credit	1.14	3.09	3.15	3.29
Energy ratio, w/ byproduct credit	1.38	5.93	6.16	3.29

[1] electricity: 12671 btu/gal
power: 36100 btu/gal   from coal natural gas   Total: 48771
Source: Gallagher and Shapouri (2006)
Also assumes   1.1 Kw-hr elec per gal eth,   3413 btu elec per Kw-hr elec, and   0.3   btu-elec per btu input

[2] harvest: 241 btu/gal
Corn Stover harvest energy from GREET model, see Wang
fertilizer replacement: 1446 btu/gal
fertilizer replacement requirement from Gallagher, Dikemen, and Shapouri (2003)
fertilizer energy requirement from GREET model, see Wang.

[3] plant, harvest, etc.   202.9 btu/gal
fertilizer   956.4 btu/gal
Plant management energy use from Heller, Keoleian, and Volk (2002).

[4] 17.6 % of baseline energy used.

distillers' grains. So neither input requires a crop production energy allocation. For corn stover, however, some fertilizer is required to replace nutrients in residues left in the field. So replacement fertilizer quantities are calculated (Gallagher et al., 2003). Then the energy embodied in the fertilizer, its production and application are calculated using the GREET model (Wang et al., 1999). Similarly, we include an allowance for external energy expended on stover harvest and machinery use. Our estimate for the fertilizer replacement and harvest, 1687 BTU/gallon in column two, is about one-third higher than the case of a willow crop. But it is still several multiples smaller than the baseline case with fossil energy.

Modification of the energy ratio for distillers' grains is straightforward. First, additional energy investment is not required because the distillers' grains are already available at the processing plant, with the external energy investment made in other parts of the corn production/processing account. So we place a 0 in the ethanol conversion entry for column 4. We also place a 0 for the by-product credit, because the distillers' grains are no longer used for a byproduct feed.

Now look at the Energy Ratio calculations for each of the power configurations. First, the baseline of column 1 is taken from a recent study (Shapouri et al., 2002). Here, the energy ratio is 1.38 with the byproduct credit, and 1.14 without byproduct credit. Without even introducing the byproduct credit, all of the biomass power alternatives have energy ratios in the 3.0 to 3.3 range. With the byproduct credit for stover and willow, the energy ratio increases by another multiple, to the 6.0 region. DG power gives a smaller improvement than other biomass alternatives when the byproduct credit is taken into account.

12.3   Crop Production and Greenhouse Gas Emissions

Corn has potential for atmospheric CO_2 reductions because it has a rapid photosynthesis reaction. Field crops are sometimes categorized according to whether they have C_3 and C_4 photosynthesis pathways. C_4 plants, which include the major ethanol feedstocks (corn and sugar), have rapid growth but require substantial nitrogen input to sustain plant growth. C_3 plants, which include soybeans, have slower plant growth and require less nitrogen (Fageria et al. 1997, pp. 44–46).

Indeed, corn replaced soybeans for the recent US ethanol expansion. So, begin by estimating and comparing net CO_2 reduction estimates for corn and beans. In Table 12.2, estimate the total biomass on an acre (column 5) using typical crop yields and biomass-grain ratios. Next, calculate the carbon content for each crop component (column 6) – carbon content estimates used composition data, and relative carbon weights for the protein (amino acid), oil (fatty acid), and starch subcomponents (White and Johnson, 2003; Yu, 2002; Kuiken and Lyman, 1948; Gallagher, 1998). Finally, applying the ratio, 3.66 lb CO_2/lb C, to the product of the carbon content estimate of column 6 and the biomass quantities in column 5, gives the CO_2/reduction estimate in column 7. This method of calculation of the CO_2 reduction is valid because photosynthesis places carbon in the plant gets there through photosynthesis. This procedure is used elsewhere (Heller et al., 2003, p. 157).

The downside is that plant growth requires fertilizer, which in turn, generates greenhouse gas emissions associated with machinery use and fertilizer application. To estimate the emission increase we again used the GREET model (Wang et al., 2007). We also confirmed

Table 12.2 Atmospheric CO2 (equivalent) reductions associated with plant growth and input use for corn and soybeans

(Col. 1) Commodity	(Col. 2) Input	(Col. 3) Crop yield (bu/acre)	(Col. 4) Grain:total biomass ratio	(Col. 5) Biomass quantity (dwt lb/acre)		(Col. 6) Carbon content (ratio)		(Col. 7) CO_2 reduction (lb CO_2/acre)
Corn		142.2	1.000	grain	6728.9	0.459		11304.2
				residue	6728.9	0.499		12289.3
				total	13457.8		total	23593.4
	Fertilizer fuel, machinery, other chemicals							−477.6
								−629.6
							net	22486.2
soybeans		28.96	0.978	grain	1563.8	0.527		3016.4
				residue	1529.4	0.577		3229.9
				total	3093.3		total	6246.3
	Fertilizer fuel, machinery, other chemicals							−92.9
								−629.6
							net	5523.8

that the GREET estimate is about the same as the default IPCC estimate for typical fertilizer application and runoff conditions in North America. In both cases, the input based emissions increases are an order of magnitude smaller than the plant based emission reduction estimates. Consequently, substantial net CO_2 reductions for an acre on both crops are calculated.

Further, the CO_2 removed from the atmosphere is reduced by a factor of 4 when 1 acre of corn replaces 1 acre of soybeans.

There is considerable discussion and some uncertainty about how unused crop residues decompose. There are N_2O emissions that the IPCC recommend estimating at 1.25 % of the nitrogen content of the crop residues. These emissions are relatively minor; we estimate the CO_2 equivalent emissions at 477.6 lbs/acre for corn and 92.9 lb/acre for soybeans.

Regarding the destination of carbon from decomposing crop residues, some presume that decomposing plant residues remain in the system, increasing carbon content of the soil, and functioning as fertilizer (Heller et al., 2003). Gallagher et al. (2003) review some data suggesting that soil carbon tends to be related to the tillage (conventional vs. no-till) method, but not to the residue practice (silage vs. field decomposition). Also, a recent simulation looks at corn residues following a switch from conventional tillage to no-till farming (Sheehan et al., 2004, p. 126). In this case, leaving the residues appears to increase soil carbon in the transition to a steady state, until about 10 years after the change in tillage practice. But there is a saturation level. Hence, removing crop residues after 10 years of no-till agricultures may not deplete soil carbon.

In subsequent analysis we assume no-till crop planting and residue removal. But residues probably shouldn't be removed prior to the saturation point to ensure sustainable production agriculture. Otherwise, residue burning would return CO_2 to the atmosphere that could have been sequestered in the soil.

12.4 CO_2 Adjustment in a Changing Ethanol Industry

The corn-ethanol industry connects a major CO_2 user, corn, to two major CO_2 producers, cars and cows. When ethanol expands, adjustments in several resource and product markets means that CO_2 balance may improve or deteriorate; the result depends on the extent of factor and product market adjustments. As the ethanol sustainability discussions move from existence to improvement issues, it is important to move beyond the conventional system boundaries of life cycle analysis. To illustrate, we consider the incremental CO_2 effects of a one gallon increase in ethanol production for two polar cases. In Table 12.3, corn supply adjusts to provide input for the ethanol industry. In Table 12.4, corn demand adjusts because the corn supply is fixed.

For estimates of the incremental CO_2 account in Table 12.3, land is diverted from soybeans to corn. The CO_2 collection estimates from Table 12.2 are used, but magnitudes are scaled to correspond to a one gallon increase in ethanol supply. Hence, the lost soybean credit (-10.85), at the bottom of the table, is the lost CO_2 collection in soybean production. But the soybean collection estimate is scaled by the amount of soybean land that must be replaced with corn to get one gallon of ethanol. The increase in corn CO_2 collection is also scaled to a one gallon increase in ethanol. Indeed, the corn credit (19.76) is derived from the carbon content of the corn used for one gallon of ethanol, less the distillers' grain

Table 12.3 Incremental CO_2 equivalent emissions budget for corn-ethanol processing and alternative processing power configurations. corn market assumption: corn land replaces soybean land; new corn is used for ethanol and DG; soybean output reduction reduces cattle feed in pounds CO_2 equivalent/gallon of ethanol, + for emission decrease and − for emission increase

	Natural gas heat & coal elec. Col. 1	Corn stover heat & elec. Col. 2	Willow heat & elec. Col. 3	Distillers' Grains Col. 4
Fuel replacement:				
corn cedit	19.76	19.76	19.76	19.76
ethanol consumption[3]	−14.12	−14.12	−14.12	−14.12
gas consumption foregone[4]	19.63	19.63	19.63	19.63
other production[2]	2.22	2.20	2.20	2.20
net	27.49	27.49	27.47	27.49
Livestock emissions change				
hogs	0	0	0	0
dairy cows	−15.88	−15.88	−15.88	0
DG credit	9.68	9.68	9.68	9.68
net	−6.20	−6.20	−6.20	9.68
Corn fertilizer				
for corn[1]	−1.25	−1.25	−1.25	−1.25
for biomass[1]	0	−0.261	−0.15	0
net	−1.25	−1.51	−1.40	−1.25
Ethanol processing energy:				
stover credit	32.06	32.06	32.06	32.06
willow credit			32.06	
stover decomposition	−32.06		−32.06	−32.06
natural gas and coal	−8.55			
stover combustion		−12.26		
willow combustion			−11.93	
DG combustion				−9.68
net	−8.55	19.80	20.13	0
Lost soybean credit	−10.85	−10.85	−10.85	−10.85
NET GAIN(+) or LOSS(−)	0.64	28.71	29.15	15.37

[1] includes machinery and fuel allowance.

[2] refinery emissions from gasoline production(3.8 lb /gal, CO_2 equi) are replaced by farm diesel, machinery, insecticide, and pesticide emissions (totaling 1.64 lb/gal, CO_2 equi).

[3] calculated using the carbon content of ethyl alcohol (52.2%) and assuming complete combustion.

[4] calculated using the carbon content of gasoline (86.3%) and assuming complete combustion.

Table 12.4 Incremental CO_2 equivalent emissions budget for corn-ethanol processing and alternative processing power configurations. corn market assumption: hog feeding reduced for all corn supply, and DG feeding to dairy cattle. No change in corn production ... in pounds CO_2 equivalent/gallon of ethanol, + for emission decrease and − for emission increase

	Natural gas heat & coal elec.	Corn stover heat & elec	Willow heat & elec.	Distillers' Grains
Fuel replacement:				
Corn credit	0	0	0	0
ethanol consumption	−14.12	−14.12	−14.12	−14.12
Gas consumption foregone	19.63	19.63	19.63	19.63
other production[2]	2.20	2.20	2.20	2.20
Net	7.71	7.71	7.71	7.71
Livestock emissions change				
hogs	12.21	12.21	12.21	12.21
dairy cattle	−20.11	−20.11	−20.11	0
DG credit	0	0	0	0
net	−7.90	−7.90	−7.90	12.21
Corn fertilizer				
for corn	−1.25	−1.25	−1.25	−1.25
for biomass[1]	0	−0.26	−0.15	0.00
net	−1.25	−1.51	−1.40	−1.25
Ethanol processing energy:				
stover credit	0	0	0	0
willow credit	0		32.06	
stover decomposition		32.06[3]		
natural gas and coal	−8.55		0	0
stover combustion		−12.26		
willow combustion			−11.93	
DG compustion				−9.68
Net	−8.55	19.80	20.13	−9.68
NET GAIN(+) or LOSS(−)	−9.99	21.10	18.54	8.99

[1]includes machinery and fuel allowance.

[2]refinery emissions from gasoline production(3.8 lb/gal, CO_2 equi) are replaced by farm deisel, machinery, insecticide, and pesticide emissions(totaling 1.64 lb/gal, CO_2 equi).

[3]decomposition forgone by using biomass for power.

byproduct, its carbon and CO_2 collection (9.68). The stover credit of 32.06 is scaled to the 1/2.7 bushels of corn. These same CO_2 collection estimates hold across all columns of Table 12.3, because the same crops are used regardless of the processing power configuration.

CO_2 emissions estimates are presented for the major uses and inputs associated with ethanol processing. The fuel replacement and livestock emissions are the same for all columns. But the input emissions change significantly across power configurations. The net fuel replacement is a gain (27.47). It consists of the CO_2 collection of corn, the ethanol fuel burning in an automobile (−14.12), the gasoline consumption foregone (19.62) and other production (2.20). Other production reflects the net gain from reduced petroleum extraction and processing against the increase in (non-fertilizer) agricultural inputs. The ethanol and gasoline estimates are calculated using the carbon content, in effect assuming an ideal engine that burns everything completely to carbon dioxide and water.

The net emissions estimate for livestock (−6.20) represents a loss. The partial credit from corn production, the carbon embodied in distillers' grains, is a gain (9.68). But the loss associated with emissions from dairy cows (−15.88) is a larger net loss. The dairy cow emission estimate used IPCC default emissions of each major greenhouse gas (carbon dioxide, methane and nitrous oxide) in North America, and conventional weighting procedures for conversion to carbon dioxide equivalents.

The corn fertilizer and ethanol processing emissions vary across columns with processing power alternatives. The fertilizer emissions include the corn crop, and in some cases, the fertilizer for the biomass crop used for ethanol energy.

Differences in emissions across power configurations are important in the net CO_2 position associated with ethanol processing:

1. The main baseline entry (col. 1) for processing emissions (−8.55) reflects the natural gas and coal used. The stover component of corn plant collections (+32.06) is a large credit, but there is a corresponding offset (−32.06) associated with stover decomposition in the atmosphere. So the net processing and the gas/power emissions are the same.
2. When corn stover is the power source (col. 2), the emissions from decomposition are replaced with a stover combustion estimate (−12.26) from GREET. Hence, net position becomes a collection instead of an emission at 19.80.
3. When willow crop is the power source (col. 3), the stover credit and decomposition are both present. Additionally, a willow credit is to account for crop growth. Finally, a woody crop combustion estimate from GREET is used again; the net processing is similar to stover, at 20.13.
4. When distillers grain is used as the power source (col. 4), an emissions estimate based on the carbon content of DGs is used. The net processing change shows as a loss, but conceptually, it offsets the DG credit from the livestock account.

Now look at the net gains and losses for an ethanol expansion at the bottom of Table 12.3. First, the Baseline net emission is a relatively small net gain of 0.62 lbs/gal, which suggests that ethanol does not improve emissions. However the net gains are considerably larger with any of the three forms of biomass power. The stover and willow cases are considerably larger, near 29 lb/gal of CO_2. The distillers' grain power improves emissions moderately, about 15 lb/gal of CO_2. The reason for the increase with all forms of biomass power is that there's an offsetting carbon collection at work with biomass power, whereas there's only an emission with the fossil fuel power in the baseline case.

Notice that the net improvement with an ethanol expansion is neutral for the baseline, even despite a significant net reduction associated with increased livestock feeding.[1] Partly, this occurs because of the positive land production credit for corn. Partly, it occurs because of our idealized perfect engine that converts all C to CO_2.

Next, consider the incremental CO_2 account of Table 12.4. Here, corn and soybean production is fixed, so the component credits (corn, DG, and stover) are all zero on the margin. Compared with Table 3S then, a level reduction in the CO_2 benefit associated with both product markets occurs. The fuel replacement net benefit is now only 7.71. The livestock net emission is slightly smaller, at −7.90, because reduced emissions from declining hog production and corn feeding, 12.21, replace the DG credit. Next, the soybean credit is excluded from Table 12.4, because there is no change in soybean production. So the baseline net (for col. 1) is −9.99, a substantial loss. Apparently, the CO_2 situation deteriorates when the ethanol industry expands by diverting corn from hog feeding when it increases dairy feeding and uses fossil fuel based power.

However, adoption of any of the biomass power options improves the net CO_2 situation regardless of whether supply or demand adjusts. For instance, the net gain improvement from switching to corn stover power is $28.71 - 0.64 = 28.07$ in Table 12.3. About the same improvement, 31.09, is obtained from Table 12.4. Further, the relative ranking of the power options is about the same for both market situations.

12.5   Conclusions

We looked at some possible changes in corn-ethanol (CE) industry practices that improve sustainability, using contributions to energy balance and global warming as the criteria. Our calculations suggest that moving from fossil fuel to biomass power can change the energy balance fraction from a moderate to a substantial contribution. Similarly adopting biomass power could induce a substantial improvement in the greenhouse gas contribution of the corn ethanol industry. On both the energy balance and global warming scores, all of the biomass power forms considered improved the situation, although some were better than others. Any or some combination of power alternatives could be included in actual implementations, after economic considerations such as production costs, and storage costs are taken into account.

Expanding the CE industry also has the potential to improve the balance of greenhouse gasses. However, there is a need to expand the traditional system boundary and incorporate

---

[1] Consider the conventional system boundary and refinery/bio-refinery comparison at the baseline. A unit of gasoline would emit:

−19.63	gasoline combustion
−3.80	refinery/extraction
−23.43	lb/total

A unit of ethanol would emit:

−14.12	ethanol combustion
−11.44	corn and ethanol production
−25.56	subtotal
+8.91	land credit, corn less soy
−16.65	

Without the land credit, ethanol emits 9.1 % more than gasoline due to higher processing emissions. With the land credit, ethanol emits 29 % less than gasoline. Wang uses a smaller land credit for corn and a smaller combustion advantage for ethanol.

LCA into economic analysis. Then realistic adjustments of agricultural land and livestock markets that accompany CE industry expansion could also be included. Our exploratory calculations suggest that the GHG balance improves when corn supply expands to accommodate increased ethanol processing. Also, the relative efficiency of corn in photosynthesis is an important contributing factor when corn-replacing-soybeans is the dominant supply adjustment. In contrast, the GHG balance deteriorates when corn demand adjusts, because the supply does not make a contribution on the margin; and because increased livestock emissions are significant.

Our exploratory calculations of incremental changes in GHG balance are a useful reference point for evaluating what happens when the CE expands. But the corn industry may need to make substantial improvements before our reference level is realized. For instance, our analysis of residue removal assumed no till farming and a decade-long adjustment period to rebuild soil carbon. Also, the nitrogen analysis assumed that the IPCC reference levels of fertilizer runoff occur. But there may also be potential for improvement far beyond the reference level. Perhaps livestock emissions can be reduced below the IPCC default levels. Alternatively, the CE industry could move to reduce the connection to the livestock industry, by using high starch corn varieties that reduce the proportion of DGs that are produced, or by using the DGs as a source of processing power.

References

Dale, Bruce E., 'Thinking Clearly about Biofuels: Ending the Irrelevant 'Net Energy' Debate and Developing Better Performance Metrics for Alternative Fuels,' *Biofuels, Bioprod. Bioref.* 1(2007): 14–17.

Fageria, N.K., V.C. Baligar, and C.A. Jones, *Growth and Mineral Nutrition of Field Crops*, 2nd edition, 1997.

Gallagher, Paul W., 'Some Productivity-Increasing and Quality Changing Technology for the Soybean Complex,' *American Journal of Agricultural Economics* 80(February 1998): 165–174.

Gallagher, P., M. Dikeman, J. Fritz, E. Wailes, W. Gauthier and H. Shapouri, 'Biomass from Crop Residues: Some Social Cost and Supply Estimates for U.S. Crops,' *Environmental and Resource Economics* 24(April 2003): 335–358.

Gallagher, Paul, Guenter Schamel, Hosein Shapouri and Heather Brubaker, 'The International Competitiveness of the U.S. Corn-Ethanol Industry: A Comparison with Sugar-Ethanol Processing in Brazil,' *Agribusiness* 22(1): 109–134 (2006).

Heller, Martin C., Gregory A. Keoleian and Timothy A. Volk, 'Life Cycle Assessment of Willow Bioenergy Cropping System,' *Biomass and Bioenergy* 25(2003): 147–165.

Intergovernmental Panel on Climate Change. Revised 1996 IPCC Guidelines for National Greenhouse Gas Inventories, 1997.

Kuiken, K.A. and Carl M. Lyman, 'Essential Amino Acid Composition of Soy Bean Meals Prepared from Twenty Strains of Soybeans,' *The Journal of Biological Chemistry*, June 11, 1948.

Pimentel, David, 'Ethanol Fuels: Energy Security, Economics, and the Environment,' *Journal of Agricultural and Environmental Ethics* 4(1991): 1–13.

Shapouri, Hosein, James A. Duffield and Michael Wang , 'The Energy Balance of Corn Ethanol: An Update,' US Department of Agriculture, Office of the Chief Economist, Office of Energy Policy and New Uses, Report No. 813 (July 2002).

Sheehan, John, Andy Aden, Keith Paustian, Kendrick Killian, John Brenner, Marie Walsh and Richard Nelson (2004) 'Energy and Environmental Aspects of Using Corn Stover for Fuel Ethanol,' *Journal of Industrial Ecology* 7(3–4): 117–146.

Wang, M., C. Sarricks, and D. Santini, 'Effects of Fuel Ethanol Use on Fuel-Cycle Energy and Green-house Gas Emissions,' Center for Transportation Research, Energy Systems Division, Argonne National Labs, Argonne, January 1999.

Wang, M., Y. Wu, and A. Elgowainy, 'The Greenhouse Gases, Regulated Emissions, and Energy Use in Transportation Model,' Operating Manual for GREET: Version 1.7, Center for Transportation Research, Argonne National Laboratory, http://www.transportaion.anl.gov/softare/GREET/index.html, February 2007.

White, P.J, and L.A. Johnson (eds), *Corn: Chemistry and Technology*, Second Edition, American Association of Cereal Chemists, Inc., St. Paul, 2003.

Yu, Guo Pei, Animal Production Based on Crop Residues – Chinese Experiences, FAO Animal Production and Health Paper 149, Food and Agriculture Organization of the United Nations, Rome 2002. ISSN 0254-6019.

Index

italic entries indicate references to a figure, **bold** entries indicate references to a table

Biofuels Edited by Wim Soetaert and Erick Vandamme
© 2009 John Wiley & Sons, Ltd